On Evaluating Curricular Effectiveness

JUDGING THE QUALITY OF K-12 MATHEMATICS EVALUATIONS

Committee for a Review of the Evaluation Data on the Effectiveness of NSF-Supported and Commercially Generated Mathematics Curriculum Materials

Mathematical Sciences Education Board
Center for Education
Division of Behavioral and Social Sciences and Education

NATIONAL RESEARCH COUNCIL
OF THE NATIONAL ACADEMIES

Program

Compliments of NPRIME
Networking Project for the Improvement of
Mathematics Education supported by the UW-System
SEA Title II Higher Education Program

THE NATIONAL ACADEMIES PRESS
Washington, D.C.
www.nap.edu

THE NATIONAL ACADEMIES PRESS 500 Fifth Street, N.W. Washington, DC 20001

NOTICE: The project that is the subject of this report was approved by the Governing Board of the National Research Council, whose members are drawn from the councils of the National Academy of Sciences, the National Academy of Engineering, and the Institute of Medicine. The members of the committee responsible for the report were chosen for their special competences and with regard for appropriate balance.

This study/publication was supported by Contract/Grant No. ESI-0102582 between the National Academy of Sciences and the National Science Foundation. Additional funding was provided by an award from the Presidents' Committee of the National Academies. Any opinions, findings, conclusions, or recommendations expressed in this publication are those of the author(s) and do not necessarily reflect the views of the organizations or agencies that provided support for the project.

Library of Congress Cataloging-in-Publication Data

On evaluating curricular effectiveness : judging the quality of K-12 mathematics evaluations / Committee for a Review of the Evaluation Data on the Effectiveness of NSF-Supported and Commercially Generated Mathematics Curriculum Materials, Mathematical Sciences Education Board, Center for Education, Division of Behavioral and Social Sciences and Education.
    p. cm.
  Includes bibliographical references.
  ISBN 0-309-09242-6 (pbk.) — ISBN 0-309-53287-6 (pdf)
  1. Mathematics—Study and teaching—Evaluation. I. National Research Council (U.S.). Committee for a Review of the Evaluation Data on the Effectiveness of NSF-Supported and Commercially Generated Mathematics Curriculum Materials.
  QA11.2.O5 2004
  510'.71—dc22
                       2004015000

Additional copies of this report are available from the National Academies Press, 500 Fifth Street, N.W., Lockbox 285, Washington, DC 20055; (800) 624-6242 or (202) 334-3313 (in the Washington metropolitan area); Internet, http://www.nap.edu.

Printed in the United States of America.

National Research Council. (2004). *On Evaluating Curricular Effectiveness: Judging the Quality of K-12 Mathematics Evaluations*. Committee for a Review of the Evaluation Data on the Effectiveness of NSF-Supported and Commercially Generated Mathematics Curriculum Materials. Mathematical Sciences Education Board, Center for Education, Division of Behavioral and Social Sciences and Education. Washington, DC: The National Academies Press.

THE NATIONAL ACADEMIES
Advisers to the Nation on Science, Engineering, and Medicine

The **National Academy of Sciences** is a private, nonprofit, self-perpetuating society of distinguished scholars engaged in scientific and engineering research, dedicated to the furtherance of science and technology and to their use for the general welfare. Upon the authority of the charter granted to it by the Congress in 1863, the Academy has a mandate that requires it to advise the federal government on scientific and technical matters. Dr. Bruce M. Alberts is president of the National Academy of Sciences.

The **National Academy of Engineering** was established in 1964, under the charter of the National Academy of Sciences, as a parallel organization of outstanding engineers. It is autonomous in its administration and in the selection of its members, sharing with the National Academy of Sciences the responsibility for advising the federal government. The National Academy of Engineering also sponsors engineering programs aimed at meeting national needs, encourages education and research, and recognizes the superior achievements of engineers. Dr. Wm. A. Wulf is president of the National Academy of Engineering.

The **Institute of Medicine** was established in 1970 by the National Academy of Sciences to secure the services of eminent members of appropriate professions in the examination of policy matters pertaining to the health of the public. The Institute acts under the responsibility given to the National Academy of Sciences by its congressional charter to be an adviser to the federal government and, upon its own initiative, to identify issues of medical care, research, and education. Dr. Harvey V. Fineberg is president of the Institute of Medicine.

The **National Research Council** was organized by the National Academy of Sciences in 1916 to associate the broad community of science and technology with the Academy's purposes of furthering knowledge and advising the federal government. Functioning in accordance with general policies determined by the Academy, the Council has become the principal operating agency of both the National Academy of Sciences and the National Academy of Engineering in providing services to the government, the public, and the scientific and engineering communities. The Council is administered jointly by both Academies and the Institute of Medicine. Dr. Bruce M. Alberts and Dr. Wm. A. Wulf are chair and vice chair, respectively, of the National Research Council.

www.national-academies.org

Foreword

The Mathematical Sciences Education Board (MSEB) of the National Research Council was established in 1985 to provide national leadership and guidance for policies, programs, and practices supporting the improvement of mathematics education at all levels. Curriculum materials for grades K-12 play a central role in what mathematics topics are taught in our schools, how the topics are sequenced and presented to students, what levels of understanding are expected, what skills students will develop and when. Schools, practitioners, policy makers, and the public depend on evaluations of materials undertaken during their development and implementation in making decisions about the appropriate uses of the materials. The MSEB recognized that the nature and quality of the evidence used to judge claims of success and failure are critical elements in enabling the community to make sound judgments. This report presents a synthesis of the evidence used in the evaluations of several sets of recently developed curriculum materials, provides a framework for the design of evaluation studies of curriculum materials, and gives conclusions and recommendations to guide future efforts in evaluating curriculum materials.

The report was prepared by a committee of experts who devoted their time, skills, and scholarship to this project over the past two years. On behalf of the MSEB, I want to thank each of them for their commitment to the important and difficult set of issues this study comprised. I especially want to commend Jere Confrey and extend deepest appreciation to her for her extraordinary leadership and commitment as chair of this project. In addition to her leadership of the committee, Jere played an extensive role in drafting and redrafting the report through the final stages of committee consultation and the intensive review process. Her dedication to maintaining the highest standards of scholarship and the full engagement of the committee, despite her many other professional obligations, was exemplary. The report bears the imprint of her commitment to intellectual and empirical rigor; the field of mathematics education is the fortunate beneficiary.

Joan R. Leitzel
Chair, MSEB

Acknowledgments

This report reflects the efforts of many people, and the committee is most grateful for their contributions. First, we wish to acknowledge and thank our sponsor, the National Science Foundation's Directorate for Education and Human Resources, for their support of this study. The National Academies' Presidents' Fund also provided some additional support for the report.

The committee was aided greatly by the individuals who participated in our evidence-gathering workshops. They helped us understand the nuances of curriculum development and implication and ways to evaluate the efficacy of such curricula once out in the field. The first evidence gathering workshop was held in September 2002 and focused on how to define or evaluate "effectiveness" of school mathematics curricula. The committee gained valuable insights from the experts who presented at that workshop: Johanna Birckmayer, Pacific Institute of Research and Evaluation; David Francis, University of Houston; Pendred Noyce, The Noyce Foundation; Andrew Porter, Wisconsin Center for Education Research; Frank Wang, Saxon Publishers; Jan Mokros, TERC; Thomas Romberg, University of Wisconsin–Madison; Sheila Sconiers, ARC Implementation Center; Andrew Isaacs, University of Chicago; Judith Zawojewski, Illinois Institute of Technology; Solomon Garfunkel, COMAP, Inc; Norman Webb, Wisconsin Center for Education Research; Richard Askey, University of Wisconsin–Madison; Roger Howe, Yale University; William McCallum, University of Arizona; M. Kathleen Heid, Pennsylvania State University; Richard Lehrer, Vanderbilt University; Richard Lesh, Purdue University; Mark Jenness,

Western Michigan University; June Mark, Education Development Center, Inc.; Maria Santos, Achieve, Inc; Richard Stanley, University of California, Berkeley; Terri Dahl, Charles M. Russell High School, MT; Eric Gutstein, University of Illinois, Chicago; Mark St. John, Inverness Research Associates; Timothy Wierenga, Naperville Community Unit School District #203, IL; James Stigler, University of California, Los Angeles; and William Tate, Washington University, St. Louis. The second evidence-gathering workshop was held in October 2002 and focused on how commercial publishers evaluate their text series. The committee gained valuable insight from Arthur Block, McGraw-Hill; Karen Usiskin, Scott Foresman; Steve Rasmussen, Key Curriculum Press; Denise McDowell, Houghton Mifflin; Paula Gustafson, Texas Education Agency; and Lisa Brady-Gill, Texas Instruments. Much was learned from those who presented at that these workshops which proved of great value to the committee as it wrote this report.

The committee would like to acknowledge the kindness of the principal investigators of the NSF-supported curriculum projects, of the NSF Implementation Centers, and of textbook publishers for supplying us with texts and evaluation studies. The committee is also most grateful to the graduate students who labored through reading and entering data on the almost 700 evaluation studies that we accumulated: Kathleen M. Clark, University of Maryland; Timothy P. Fukawa-Connelly, University of Maryland; Michelle Caren Massey, American University; and Francesca Palik, George Washington University. Among these graduate students, special thanks go to Erica Slate Young, University of Texas, Austin, who worked closely with the chair of the committee. Sibel Kazak and Dustin Mitchell, Washington University in St. Louis also provided assistance to the chair. In addition special thanks go to Alan Maloney, for the long hours he dedicated (including his birthday) to helping the chair prepare the final revised draft of the report.

Special thanks go to the NRC staff who worked on this report. Research associate Vicki Stohl worked tirelessly to develop a data base, assemble materials for the committee, and produce part of the preliminary draft of the document. Senior program assistant Dionna Williams worked efficiently under great time pressure to engineer the workshops and other committee meetings and to put the many drafts of this report into good form. Carole Lacampagne provided leadership for the project, Jay Labov gave sage advice along the way, and Patricia Morison was instrumental in bringing this report to successful closure. Martin Orland, director of the Center for Education, also provided leadership and support for the project.

This report has been reviewed in draft form by individuals chosen for their diverse perspectives and technical expertise, in accordance with procedures approved by the NRC's RRC. The purpose of this independent review is to provide candid and critical comments that will assist the institution in making its published report as sound as possible and to ensure that the report meets institutional standards for objectivity, evidence, and responsiveness to the study charge. The review comments and draft manuscript remain confidential to protect the integrity of the deliberative process. We wish to thank the following individuals for their review of this report: Douglas Carnine, National Center to Improve the Tools of Educators, University of Oregon; Mary M. Lindquist, Department of Mathematics Emeritus, Columbus State University, Columbus, GA; R. James Milgram, Department of Mathematics, Stanford University; Leona Schauble, Teaching and Learning Department, Vanderbilt University; Alan H. Schoenfeld, School of Education, University of California, Berkeley; Norman Webb, Wisconsin Center for Education Research, University of Wisconsin–Madison; and Iris R. Weiss, President, Horizon Research, Inc., Chapel Hill, NC.

Although the reviewers listed above have provided many constructive comments and suggestions, they were not asked to endorse the conclusions or recommendations nor did they see the final draft of the report before its release. The review of this report was overseen by Adam Gamoran, University of Wisconsin–Madison, and William G. Howard, Jr., Independent Consultant, Scottsdale, AZ, appointed by the National Research Council, who were responsible for making certain that an independent examination of this report was carried out in accordance with institutional procedures and that all review comments were carefully considered. Responsibility of the final content of this report, however, rests entirely with the authoring committee and the NRC.

Contents

On
Evaluating
Curricular
Effectiveness

Executive Summary

Curricula play a vital role in educational practice. They provide a crucial link between standards and accountability measures. They shape and are shaped by the professionals who teach with them. Typically, they also determine the content of the subjects being taught. Furthermore, because decisions about curricula are typically made at the local level in the United States, a wide variety of curricula are available for any given subject area. Clearly, knowing how effective a particular curriculum is, and for whom and under what conditions it is effective, represents a valuable and irreplaceable source of information to decision makers, whether they are classroom teachers, parents, district curriculum specialists, school boards, state adoption boards, curriculum writers and evaluators, or national policy makers. Evaluation studies can provide that information but only if those evaluations meet standards of quality.

Under the auspices of the National Research Council, this committee's charge was to evaluate the quality of the evaluations of the 13 mathematics curriculum materials supported by the National Science Foundation (NSF) (an estimated $93 million) and 6 of the commercially generated mathematics curriculum materials (listing in Chapter 2).

The committee was charged to determine whether the currently available data are sufficient for evaluating the effectiveness of these materials and, if these data are not sufficiently robust, the committee was asked to develop recommendations about the design of a subsequent project that could result in the generation of more reliable and valid data for evaluating these materials.

The committee emphasizes that it was not charged with and therefore did not:

- Evaluate the curriculum materials directly; or
- Rate or rank specific curricular programs.

In addressing its charge, the committee held fast to a single commitment: that our greatest contribution would be to clarify the proper elements of an array of evaluation studies designed to judge the effectiveness of mathematics curricula and clarify what standards of evidence would need to be met to draw conclusions on effectiveness.

ASSESSMENT OF EXISTING STUDIES

The committee began by systematically identifying and examining the large array of evaluation studies available on these 19 curricula. In all, 698 studies were found. The first step in our process was to eliminate studies that were clearly not evaluations of effectiveness—those lacking relevance or adequacy for the task (e.g., product descriptions, editorials) (n=281), and those classified as providing background information, historical perspective, or a project update (n=225). We then categorized the remaining (192) studies into the four major evaluation methodologies—content analyses (n=36), comparative studies (n=95), case studies (n=45), and syntheses (n=16). Criteria by which to judge methodological adequacy, specific to each methodology, were then used to decide whether studies should be retained for further examination by the committee.

Content analyses focus almost exclusively on examining the content of curriculum materials; these analyses usually rely on expert review and judgments about such things as accuracy, depth of coverage, or the logical sequencing of topics. For the 36 studies classified as content analyses, the committee drew on the perspectives of eight prominent mathematicians and mathematics educators, in addition to applying the criteria of requiring full reviews of at least one year of curricular material. All 36 studies of this type were retained for further analysis by the committee.

Comparative studies involve the selection of pertinent variables on which to compare two or more curricula and their effects on student learning over significant time periods. For the 95 comparative studies, the committee stipulated that they had to be "at least minimally methodologically adequate," which required that a study:

- Include quantifiably measurable outcomes such as test scores, responses to specified cognitive tasks of mathematical reasoning, performance evaluations, grades, and subsequent course taking; and

- Provide adequate information to judge the comparability of samples.

In addition, a study must have included at least one of the following additional design elements:

- A report of implementation fidelity or professional development activity;
- Results disaggregated by content strands or by performance by student subgroups; or
- Multiple outcome measures or precise theoretical analysis of a measured construct, such as number sense, proof, or proportional reasoning.

The application of these criteria led to the elimination of 32 comparative studies.

Case studies focus on documenting how program theories and components of a particular curriculum play out in a particular real-life situation. These studies usually describe in detail the large number of factors that influence implementation of that curriculum in classrooms or schools. For the 45 case studies, 13 were eliminated leaving 32 that met our standards of methodological rigor.

Synthesis studies summarize several evaluation studies across a particular curriculum, discuss the results, and draw conclusions based on the data and discussion. All of the 16 synthesis studies were retained for further examination by the committee.

The committee then had a total of 147 studies that met our minimal criteria for consideration of effectiveness, barely more than 20 percent of the total number of submissions with which we began our work. Seventy-five percent of these studies were related to the curricula supported by the National Science Foundation. The remaining studies concerned commercially supported curricular materials.

On the basis of the committee's analysis of these 147 studies, we concluded that the corpus of evaluation studies as a whole across the 19 programs studied does not permit one to determine the effectiveness of individual programs with a high degree of certainty, due to the restricted number of studies for any particular curriculum, limitations in the array of methods used, and the uneven quality of the studies.

This inconclusive finding should not be interpreted to mean that these curricula are not effective, but rather that problems with the data and/or study designs prevent confident judgments about their effectiveness. Inconclusive findings such as these do not permit one to determine conclusively whether the programs overall are effective or ineffective.

A FRAMEWORK FOR FUTURE EVALUATIONS

Given this conclusion, the committee turned to the second part of its charge, developing recommendations for future evaluation studies. To do so, the committee developed a framework for evaluating curricular effectiveness. It permitted the committee to compare evaluations and consider how to identify and distinguish among the variety of methodologies employed.

The committee recommends that individuals or teams charged with curriculum evaluations make use of this framework. The framework has three major components that should be examined in each curriculum evaluation: (1) the program materials and design principles; (2) the quality, extent, and means of curricular implementation; and (3) the quality, breadth, type, and distribution of outcomes of student learning over time.

The quality of an evaluation depends on how well it connects these components into a research design and measurement of constructs and carries out a chain of reasoning, evidence, and argument to show the effects of curricular use.

ESTABLISHING CURRICULAR EFFECTIVENESS

The committee distinguished two different aspects of determining curricular effectiveness. First, each individual study should demonstrate that it has obtained a level of *scientific validity*. In the committee's view, for a study to be *scientifically valid,* it should address the components identified in the framework and it should conform to the methodological expectations of the appropriate category of evaluation as discussed in the report (content analysis, comparative study, or case study).

Defining scientific validity for individual studies is an essential element of assuring valid data about curricular effectiveness. However, curricular effectiveness cannot be established by a single scientifically valid study; instead a body of studies is needed, which is the second key aspect of determining effectiveness. Curricular effectiveness is an integrated judgment based on interpretation of a number of scientifically valid evaluations that combine social values, empirical evidence, and theoretical rationales.

Furthermore, a single methodology, even replications and variations of a study, is inadequate to establish curricular effectiveness, because some types of critical information will be lacking. For example, a content analysis is important because, through expert review of the curriculum content, it provides evidence about such things as the quality of the learning goals or topics that might be missing in a particular curriculum. But it cannot determine whether that curriculum, when actually implemented in classrooms, achieves better outcomes for students. In contrast, a comparative study can provide evidence of improvement in student learning in real classrooms across different curricula. Yet without the kind of complementary evidence provided in a content analysis, nothing will be known about the quality or comprehensiveness of the content in the curriculum that produced better outcomes. Furthermore, neither content analyses nor comparative studies typically provide information about the quality of the implementation of a particular curriculum. A case study provides deep insight into issues of implementation; by itself, though, it cannot establish representativeness or causality.

This conclusion—that multiple methods of evaluation strengthen the determination of effectiveness—led the committee to recommend that a curricular program's effectiveness should be ascertained through the use of multiple methods of evaluation, each of which is a scientifically valid study. Periodic synthesis of the results across evaluation studies should also be conducted.

This is a general principle for the conduct of evaluations in recognition that curricular effectiveness is an integrated judgment, continually evolving, and based on scientifically valid evaluations. The committee further recognized, however, that agencies, curriculum developers, and evaluators need an explicit standard by which to decide when federally funded curricula (or curricula from other sources whose adoption and use may be supported by federal monies) can be considered effective enough to adopt. The committee proposes a rigorous standard to which programs should be held to be *scientifically established as effective*.

In this standard, the committee recommends that a curricular program be designated as *scientifically established as effective* only when it includes a collection of scientifically valid evaluation studies addressing its effectiveness that establish that an implemented curricular program produces valid improvements in learning for students, and when it can convincingly demonstrate that these improvements are due to the curricular intervention. The collection of studies should use a combination of methodologies that meet these specified criteria: (1) content analyses by at least two qualified experts (a Ph.D.-level mathematical scientist and a Ph.D.-level mathematics educator) (required); (2) comparative studies using experimental or quasi-experimental designs, identifying the comparative curriculum (required); (3) one or more case studies to investigate the relationships among the implementation of the curricular program and the program components (highly desirable); and (4) a final report, to be made publicly available, should link the analyses, specify what they convey about the effectiveness of the curriculum, and stipulate the extent to which the program's effectiveness can be generalized (required). This standard relies on the primary methodologies identified in our review, but we acknowledge the possibility of other configurations, provided they draw on the framework and the definition of scientifically valid studies and include careful review and synthesis of existing evaluations.

In its review, the committee became concerned about the lack of independence of some of the evaluators conducting the studies; in too many cases, individuals who developed a particular curriculum were also members of the evaluation team, which raised questions about the credibility of the evaluation results. Thus, to ensure the independence and impartiality of evaluations of effectiveness, the committee also recommends that summative evaluations be conducted by independent evaluation teams with no membership by authors of the curriculum materials or persons under their supervision.

In the body of this report, the committee offers additional recommended practices for evaluators, which include:

Representativeness. Evaluations of curricular effectiveness should be conducted with students that represent the appropriate sampling of all intended audiences.

Documentation of implementation. Evaluations should present evidence that provides reliable and valid indicators of the extent, quality, and type of the implementation of the materials. At a minimum, there should be documentation of the extent of coverage of curricular material (what some investigators referred to as "opportunity to learn") and of the extent and type of professional development provided.

Curricular validity of measures. **A minimum of** one of the outcome measures used to determine curricular effectiveness should possess demonstrated curricular validity. It should comprehensively sample the curricular objectives in the course, validly measure the content within those objectives, ensure that teaching to the test (rather than the curriculum) is not feasible or likely to confound the results, and be sensitive to curricular changes.

Multiple student outcome measures. Multiple forms of student outcomes should be used to assess the effectiveness of a curricular program. Measures should consider persistence in course taking, drop-out or failure rates, as well as multiple measures of a variety of the cognitive skills and concepts associated with mathematics learning.

Furthermore, the committee offers recommendations about how to strengthen each of the three major curriculum evaluation methodologies.

Content analyses. A content analysis should clearly indicate the extent to which it addresses the following three dimensions:

1. Clarity, comprehensiveness, accuracy, depth of mathematical inquiry and mathematical reasoning, organization, and balance (disciplinary perspectives).

2. Engagement, timeliness and support for diversity, and assessment (learner-oriented perspectives).

3. Pedagogy, resources, and professional development (teacher- and resource-oriented perspectives).

In considering these dimensions, specific evidence of each should be provided to support their judgments. A content analysis should be acknowledged as a connoisseurial assessment and should include identified credentials and statements of preference and bias of the evaluators.

Comparative analyses. As a result of our study of the set of 63 *at least minimally methodologically adequate* comparative analyses, the committee recommends that in the conduct of all comparative studies, explicit attention be given to the following criteria:

- Identify comparative curricula by name;
- Employ random assignment, or otherwise establish adequate comparability;
- Select the appropriate unit of analysis;
- Document extent of implementation fidelity;
- Select outcome measures that can be disaggregated by content strand;
- Conduct appropriate statistical tests and report effect size;
- Disaggregate data by gender, race/ethnicity, socioeconomic status (SES), and performance levels, and express constraints as to the generalizability of study.

The committee recognized the need to strengthen the conduct of comparative studies in relation to the criteria listed above. It also recognized that much could be learned from the subgroup (n=63) identified as "at least minimally methodologically adequate." In fields in their infancy, evaluators and researchers must pry apart issues of method from patterns of results. Such a process requires one to subject the studies to alternative interpretation; to test results for sensitivity or robustness to changes in design; to tease out among the myriad of variables, the ones most likely to produce, interfere with, suppress, modify, and interact with the outcomes; and to build on results of previous studies. To fulfill the charge to inform the conduct of future studies, in Chapter 5 the committee designed and conducted methods to test the patterns of results under varying conditions, and to determine which patterns were persistent or ephemeral. We used these analyses as a baseline to investigate the question, Does the application of increasing standards of rigor have a systemic effect on the results?

In doing so, we report the patterns of results separately for evaluations of NSF-supported and commercially generated programs because NSF-supported programs had a common set of design specifications including consistency with the National Council of Teachers of Mathematics (NCTM) Standards, reliance on manipulatives, drawing topics from statistics, geometry, algebra and functions, and discrete mathematics at each grade level, and strong use of calculators and computers. The commercially supported curricula sampled in our studies varied in their use of these curricular approaches; further subdivisions of these evaluations are also presented in the report. The differences in the specifications of the two groups of programs make their evaluative procedures and hence the validation of those procedures so unlike each other, that combining them into a single category could be misleading.

One approach taken was to filter studies by separating those that met a particular criterion of rigor from those that did not, and to study the effects of that filter on the pattern of results as quantified across outcome measures into the proportion of findings that were positive, negative, or indeterminate (no significant difference). First, we found that on average the evaluations of the NSF-supported curricula (n=46) in this subgroup had reported stronger patterns of outcomes in favor of the experimental curricula than had the evaluations of commercially generated curricula (n=17). Again we emphasize that due to our call for increased methodological rigor and the use of multiple methods, this result is not sufficient to establish the curricular effectiveness of these programs as a whole with adequate certainty. However, this result does provide a testable hypothesis, a starting point for others to examine, critique, and undertake further studies to confirm or disconfirm. Then, after applying the criteria listed above, we found that the comparative studies of both NSF-supported and commercially generated curricula that had used the more rigorous criteria never produced contrary conclusions about curricular effectiveness (compared with less rigorous methods). Furthermore, when the use of more rigorous criteria did lead to significantly different results, these results tended to show weaker findings about curricular effects on student learning. Hence, this investigation reinforced the importance of methodological rigor in drawing appropriate inferences of curricular effectiveness.

Case studies. Case studies should meet the following criteria:

- Stipulate clearly what they are cases of, how claims are produced and backed by evidence, and what events are related or left out and why; and
- Identify explicit underlying mechanisms to explain a rich variety of research evidence.

The case studies should provide documentation that the implementation and outcomes of the program are closely aligned and consistent with the curricular program components and add to the trustworthiness of implementation and to the comprehensiveness and validity of the outcome measures.

The committee recognizes the value of diverse curricular options and finds continuing experimentation in curriculum development to be essential, especially in light of changes in the conduct and use of mathematics and technology. However, it should be accompanied by rigorous efforts to improve our conduct of evaluation studies, strengthening the results by learning from previous efforts.

RECOMMENDATIONS TO FEDERAL AGENCIES, STATE AND LOCAL DISTRICTS AND SCHOOLS, AND PUBLISHERS

Responsibility for curricular evaluation is shared among three primary bodies: the federal agencies that develop curricula, publishers, and state and local districts and schools. All three bodies can and should use the framework and guidelines in designing evaluation programs, sponsoring appropriate data collections, reviewing evaluation proposals, and assessing evaluation studies. The committee has identified several short- and long-term actions that these bodies can take to do so.

At the federal level, such actions include:

- Specifying more explicit expectations in requests for proposals for evaluation of curricular initiatives and increasing sophistication in methodological choices and quality;
- Denying continued funding for major curricular programs that fail to present evaluation data from well-designed, scientifically valid studies;
- Charging a federal agency with responsibility for collecting and maintaining district- and school-level data on curricula; and
- Providing training, in concert with state agencies, to district and local agencies on conducting and interpreting studies of curricular effectiveness.

For publishers, such actions include:

- Differentiating market research from scientifically valid evaluation studies; and
- Making evaluation data available to potential clients who use federal funds to purchase curriculum materials.

At the state level, such actions include:

- Developing resources for district- and state-level collection and maintenance of data on issues of curricular implementation; and

- Providing districts with training on how to conduct feasible, cost-efficient, and scientifically valid studies of curricular effectiveness.

At the district and local levels, such actions include:

- Improving methods of documenting curricular use and linking it to student outcomes;
- Maintaining careful records of teachers' professional development activities related to curricula and content learning; and
- Systematically ensuring that all study participants have had fair opportunities to learn sufficient curricular units, especially under conditions of student mobility.

Finally, the committee believes there is a need for multidisciplinary basic empirical research studies on curricular effectiveness. The federal government and publishers should support such studies on topics including, but not limited to:

- The development of outcome measures at the upper level of secondary education and at the elementary level in non-numeration topics that are valid and precise at the topic level;
- The interplay among curricular implementation, professional development, and the forms of support and professional interaction among teachers and administrators at the school level;
- Methods of observing and documenting the type and quality of instruction;
- Methods of parent and community education and involvement, and
- Targets of curricular controversy such as the appropriate uses of technology; the relative use of analytic, visual, and numeric approaches; or the integration or segregation of the treatment of subfields, such as algebra, geometry, statistics, and others.

The committee recognizes the complexity and urgency of the challenge the nation faces in establishing effectiveness of mathematics curricula, and argues that we should avoid seemingly attractive, but oversimplified, solutions. Although the corpus of evaluation studies is not sufficient to directly resolve the debates on curricular effectiveness, we believe that in the controversy surrounding mathematics curriculum evaluation, there is an opportunity. This opportunity should not be missed to forge solutions through negotiation of perspective, to base our arguments on empirical data informed by theoretical clarity and careful articulation of values, and to build in an often-missing measure of coherence to curricular choice, and feedback from careful, valid, and rigorous study. Our intention in presenting this report is to help take advantage of that opportunity.

1

The Challenge and the Charge

In the United States, where much educational decision making is undertaken at the state or local level, the availability of a variety of curricula is both expected and desired. However, the many products and approaches to curricula are likely to result in varied quality and effectiveness. Consequently, state and local decision makers need valid, informative, credible, and cost-efficient evaluation data on curricula effectiveness to assist them in the interpretation and use of these data. National-level policy makers and agencies and commercial publishers that support the development of curricula also must be assured that the funds expended for such purposes result in development of curricula and associated resources that demonstrably enhance learning. Methodologically sound evaluations of those materials are essential.

However, no single method of evaluation alone is sufficient. Evaluation necessarily involves value judgments and requires careful consideration of evidence. Well-conducted evaluations depend on the availability and distribution of resources, are expensive to undertake, and reflect contextual opportunities and constraints. Thus, decision makers need a flexible evaluation framework that provides a highly reliable and informative means of curricular review that fits local goals and expectations. Moreover, curricular decisions must be reexamined periodically, and curricula need to be revised based on data and professional judgment. Curriculum evaluations must accommodate local expectations, values, and resources.

To address this issue, a committee (hereafter referred to as "we") was assembled by the National Research Council (NRC) in spring 2002. Our assignment was to collect the evaluation studies of certain mathematics curricula developed by for-profit companies or with National Science Foundation (NSF) funds, or by a combination of the two, and to assess their quality. This report presents our conclusions and provides recommendations for improvements to the evaluation process.

NEED FOR THIS STUDY

Between 1990 and 2007, the NSF will have devoted an estimated $93 million, including funding for revisions, to 13 mathematics projects to "stimulate the development of exemplary educational models and materials (incorporating the most recent advances in subject matter, research in teaching and learning, and instructional technology) and facilitate their use in the schools" (NSF, 1989, p. 1). As these NSF-supported materials, which were informed by the publication of the National Council of Teachers of Mathematics (NCTM) Standards (NCTM, 1989), gained visibility, publishers also produced curriculum materials aligned with NCTM Standards or developed alternative approaches based on other standards.

These standards were viewed as a promising new approach for translating and infusing research results into classroom practice across the United States. Although each NSF-supported curriculum underwent individual evaluations, little emphasis was placed on reaching consensus about the particular aspects of the curricula to be analyzed or methods to be used. Furthermore, until these curricula had been used for a significant amount of time, no meta-analysis of NSF efforts as a whole in supporting new mathematics curricula could be undertaken.

In 1999, the U.S. Department of Education convened a Panel on Exemplary Programs in Mathematics whose recommended curriculum programs generated much controversy (Klein et al., 1999). Documented evidence of a curriculum's effectiveness was included in the Panel's criteria. Part of the controversy concerned the quality of this evidence. Because the NSF-supported materials have been marketed longer and additional evaluation studies have been conducted, reexamination of the adequacy of the evaluations is timely.

Such examination is essential because several factors indicate that the conditions that motivated NSF funding of those curriculum projects may still persist (McKnight et al., 1987; Schmidt, McKnight, and Raizen, 1996). The United States may not be meeting its own mathematical needs in producing students who are capable, interested, and successful in the following areas:

- Attaining high school diplomas with adequate levels of mathematical knowledge and reasoning to function as an informed and critical citizenry (Adelman, 1999);
- Undertaking study at two-year colleges without undue fiscal burdens imposed by the need for remedial mathematics activities (Adelman, 1999);
- Pursuing advanced mathematics at the research level in mathematics and science (Lutzer, 2003); and
- Pursuing mathematically intensive careers in technology fields, statistics, and "client disciplines"—engineering, chemistry, and, increasingly, fields such as biology, economics, and social sciences (NRC, 2003).

In addition, concerns for preparation of all students (Campbell et al., 2002) across the spectrum of academic achievement necessitate such examination, evaluation, and critique of mathematics curricula.

Currently, too many deliberations on mathematics curricular choices lack a careful and thorough review of the evaluations of mathematics curricula. Because of the cumulative nature of mathematics topics, a weak curriculum can limit and constrain instruction beyond the K-12 years. It can discourage students from entering mathematically intensive fields or hobble the progress of those who pursue them. International studies have heightened American awareness that our mathematics performance has deteriorated, especially in the 8th and 12th grades. Even the performance of the most advanced students has suffered (Takahira et al., 1998).

The impetus for ways to examine effectiveness of curricular reform was intensified with release of the 2003 National Assessment of Educational Progress report, known as the Nation's Report Card, which showed significant improvements in mathematics achievement as reading scores remained constant. Average 4th-grade student performance increased nine points, while 8th-grade student scores increased by five points. Closer examination shows that the percentage of students identified as below basic levels of performance declined by 12 and 5 percentage points at the 4th and 8th grades, respectively. The majority of subsequent gains occurred in the number of students identified as proficient, the second-highest level. These gains were quite evenly distributed across ethnic groups and class lines. Interpreting the scores over successive years created methodological issues, and the factors instrumental in producing these gains are not known. Determining the extent to which these gains can be attributed to curricular reform requires application of sound, sophisticated evaluation design, establishing an additional need for this report.

TIMELINESS OF THE REPORT

This report is timely because a review of evaluations providing evidence on the effectiveness of mathematics curricula must be undertaken after the curriculum materials have been used under a variety of conditions and when the materials are in final editions rather than preliminary forms. Premature review would contribute to unrealistic perceptions that education can be easily fixed in a short period. An early review also could contribute to vacillation among approaches, wasted funding, and practitioners skeptical of change who cringe as they await future reforms to displace current efforts.

This review is also timely because of the federal No Child Left Behind Act of 2001. This law specifies that all educational programs should demonstrate effectiveness based on "scientifically based research." Publishers, decision makers, and researchers are now seeking clear guidelines to determine whether their curriculum development programs meet this standard. Guidelines must be designed that are informed by and built on the state of evaluation data currently available. As committee members, we believe that funding decisions should be predicated on a realistic, honest assessment of the quality of the current knowledge base. Given this legislative mandate, we sought to **define** the phrase *scientifically established as effective* as applied to mathematics curricula. Our deliberations also have been informed by the use of the phrase *scientific research in education,* as articulated by the NRC report with the same name (NRC, 2002).

COMMITTEE CHARGE AND APPROACH

Our committee was assembled in June 2002 with the following charge:

The Mathematical Science Education Board will nominate a committee of experts in mathematics assessment, curriculum development, curriculum implementation, and teaching to assess the quality of studies about the effectiveness of 13 sets of mathematics curriculum materials developed through NSF support and 6 sets of commercially generated curriculum materials. A committee will collect available studies that have evaluated NSF-supported development and commercially generated mathematics education materials and establish initial criteria for review of the quality of those studies. The committee will receive input from two workshops of mathematics educators, mathematicians, curriculum developers, curriculum specialists, and teachers. The product will be a consensus report to NSF summarizing the results of the workshops, presenting the criteria and framework for reviewing the evidence, and indicating whether the currently available data are sufficient for evaluating the efficacy of these materials. If these data are not sufficiently robust, then the steering committee would also develop recommendations about the design of a subsequent project that could result in the generation of more reliable and valid data for evaluating these materials.

Originally we were to review evaluation data on the effectiveness of only NSF-supported mathematics curriculum materials. Our charge was amended to include evaluation data on the effectiveness of six sets of commercially generated mathematics materials. This expanded scope anticipated that methods of evaluation and data thus derived from commercially generated materials might differ from the methods used to evaluate the NSF-supported curriculum materials. By expanding its investigation to include commercially generated mathematics curricula, we anticipated learning about different techniques that might be incorporated into a curriculum evaluation framework. Investigating these alternative approaches to evaluation might be useful to a broader spectrum of people who evaluate mathematics curricula.

Our goal in writing this report is twofold. First, we aim to examine evidence currently available from the evaluation of effectiveness of mathematics curricula. Second, we will suggest ways to improve the evaluation process that will enhance the quality and usefulness of evaluations and help guide curriculum developers and evaluators in conducting better studies. To determine if the corpus of evaluations was "sufficient for reviewing the efficacy of the materials," we examined both their methods and the conclusions, evaluating the quality of evidence and argument. We also distinguished between studies that were at least minimally methodologically adequate and those with methodology lacking sufficient rigor or relevance. Finally, "in order to make recommendations about the design of a subsequent project," we summarized inferences that could be drawn from the patterns of findings of those "at least minimally methodologically adequate" studies that would inform the design and conduct of subsequent evaluations and an evaluation framework. However, to stay within the limits of our charge, we do not report summary data at the level of particular programs. Instead, we report at the level of program type, and use the summary data as a means to investigate the quality and stability of the evaluations. Furthermore, we recognize that design weaknesses of some evaluation studies render the summary statements only tentative. In this way, we sought to fulfill our charge by advancing "the design of a subsequent project that could result in the generation of more reliable and valid data for evaluating these materials."

Establishing clearer guidelines for curricular evaluation becomes increasingly important as the number of U.S. publishing companies decreases through mergers, acquisitions, and purchase by international publishing conglomerates. This reduction in publishing companies is likely to affect curriculum development, review, revision, and adoption. Also needed are criteria that enable researchers and policy makers to monitor and document the effects of these changes on future curricular options that become available to U.S. educators and students. Members of these corporations indicated to us that they welcome clear statements of their responsibilities in this arena of curricula evaluation.

Thus, this report considers issues related to policy, practice, and measurement in an integrated fashion. Policy makers should be knowledgeable of real practice demands and their effect on evaluations. They need expert advice on how to develop a plan of action that serves the needs of all constituents and is reliable, strategic, and feasible. At the same time, practice in education is complex and subject to multiple forces. It exists within multiple levels of organization, governance, and regulation. Practitioners, the majority of whom are teachers of mathematics, are charged with mathematics curricular implementation, and their professional preparation, knowledge, and experience are essential in selecting materials for their curricular effectiveness. Curricular evaluation must consider not only the quality of the materials but also a realistic assessment of the practice conditions in which these innovations are set. Thus, our efforts address the intended curriculum and the enacted curriculum.[1] Finally, undertaking these studies within a scientific approach to educational research requires the clear articulation of the tenets that underlie evaluations of curricula effectiveness.

REPORT LAYOUT

Chapter 2 begins with a discussion of the methods used to collect relevant evaluation studies. It describes the resulting database, methods, and criteria used to review these studies and to decide which evaluation studies should be included in the report. This chapter also describes the initial study characteristic coding system that was used to create and analyze the large database.

The database and study characteristics were then used to develop a framework for curriculum evaluation in mathematics. This framework is presented in Chapter 3. Based on the framework, we identified four major classes of evaluation studies—content analysis, comparative analysis, case, and synthesis. We divided into four subgroups to study each in depth. The subgroups refined the methodology to create a decision tree to map studies

---

[1]Intended curriculum is the subject matter, skills, and values that policy makers or developers expect to be taught and enacted curriculum is the curriculum that was implemented in the classroom (Goodlad, 1984; Cuban, 1992).

into categories for further examination. Discussions of each of these categories, together with the refined methodology, appear in three chapters: Chapter 4 details content analysis studies, Chapter 5 details comparative studies, and Chapter 6 details case and synthesis studies. These subgroup reports were subsequently reviewed and discussed by the entire committee and were revised to relate to each other and to the framework.

Our conclusions and recommendations are listed in Chapter 7.

2

Committee Procedures and Characteristics of Studies Reviewed

As explained in Chapter 1, our charge as committee members was to evaluate the quality of the evaluations of the 13 National Science Foundation-(NSF-) supported and 6 commercially generated mathematics curriculum materials.

We were not charged with and therefore did not:

- **Evaluate the curriculum materials directly**
- **Rate or rank specific curricular programs**

We recognize that both tasks could interest a broad constituency, but we believed that the field would profit from a careful, thorough review and summary of previous evaluations and research studies in relation to how previous work might inform and strengthen future efforts. We were aware that the mathematics education field lacks a clear consensus on what constitutes an effective curriculum and how to measure it to provide adequate, valid, and timely information to *decision making* bodies. It is appropriate to have a range of curricula from which to choose that represent a variety of preferences and values; when this is the case, decision making on curricular materials inevitably combines values and evidence-based reasoning. We did not intend to recommend the elimination of values in curricular decision making, but instead wished to contribute to efforts to increase the quality of evidence provided to the process.

Some readers may be disappointed by our not offering a "stamp of approval" on specific curricula or providing a report card, as others have done for state standards or tests (U.S. Department of Education, 1999; Achieve Inc., 2002). This decision was deliberate. As a committee of the National Research Council of The National Academies, our primary contribution was to clarify the phrases *scientifically valid evaluation study* and *scientifically established as effective* in the context of K-12 mathematics curricula. Such an analysis can elucidate the current knowledge of how these curricula were evaluated and help decision makers avoid judgment errors that are likely when the completeness or scientific rigor of evaluations of such materials is misunderstood.

Recognizing the complexity of judging curricular effectiveness, this report is designed to assist future evaluators and policy makers in designing and conducting evaluations that provide accurate, comprehensive, and valid advice to decision makers and practitioners on the efficacy of curriculum materials. Our primary goal was to advise our audiences on what could be learned from these initial efforts and how lessons learned, strategic decisions, adaptations in method, errors and weaknesses, and tentative patterns of results could further future evaluation efforts and decision making on curricular policy.

CURRICULA UNDER REVIEW

The following 13 mathematics curricula programs[1] (The K-12 Mathematics Curriculum Center, 2002) were supported by the NSF, and the evaluations of these materials were reviewed by our committee:

Elementary School:

- *Everyday Mathematics (EM), Grades K-6* (SRA/McGraw-Hill)
- *Investigations in Number, Data and Space, Grades K-6* (Scott Foresman)
- *Math Trailblazers, Grades K-6* (Kendall/Hunt Publishing Company)

Middle School:

- *Connected Mathematics Project (CMP), Grades 6-8* (Prentice Hall)

---

[1]Each of the NSF-supported curricula is at least a three-year core curriculum (National Science Foundation, 1989, 1991). A condition of second-year funding for the NSF-supported curricula materials was a firm commitment by a publisher for national dissemination (National Science Foundation, 1989, 1991).

- *Mathematics in Context (MiC), Grades 5-8* (Holt, Rinehart and Winston)
- *MathScape: Seeing and Thinking Mathematically, Grades 6-8* (Glencoe/McGraw-Hill)
- *MathThematics (STEM), Grades 6-8* (McDougal Littell)
- *Middle School Mathematics Through Applications Project (MMAP) Pathways to Algebra and Geometry, Grades 6-8* (currently unpublished)

High School:

- *Contemporary Mathematics in Context (Core-Plus), Grades 9-12* (Glencoe/McGraw-Hill)
- *Interactive Mathematics Program (IMP), Grades 9-12* (Key Curriculum Press)
- *MATH Connections: A Secondary Mathematics Core Curriculum, Grades 9-12* (IT'S ABOUT TIME, Inc.)
- *Mathematics: Modeling Our World (MMOW/ARISE), Grades 9-12* (W.H. Freeman and Company)
- *Systemic Initiative for Montana Mathematics and Science (SIMMS) Integrated Mathematics, Grades 9-12* (Kendall/Hunt Publishing Company)

Given our expanded charge, we also included a few of the commercially published, non–NSF-funded curricula. We planned to select the curricula by market share; however, such data are highly proprietary and contested. An additional complicating factor was that most reports of market share are identified by publisher name rather than a particular product line. Publishers produce numerous overlapping and sometimes competing mathematics curriculum products, especially given recent acquisitions and mergers. Thus determining market share by program was problematic.

We located two sources of market share data independent of the publishers (Education Market Research, 2001; Weiss et al., 2001). In addition, we received testimonial data from other suppliers of widely used curricular materials in mathematics, including Key Curriculum Press, Saxon Publishers,[2] and Texas Instruments. Among the six curricula, we sought representation from each of the four major textbook publishers:

---

[2] Saxon Publishers suggested Simba Information Inc.'s (2002, 2003) *Print Publishing for the School Market 2001-2002 Yearly Report* and Educational Marketer's monthly newsletter as sources for market share data.

1. McGraw-Hill (including Direct Instruction, Frank Schaffler Publishing, Macmillan, Glencoe, SRA/Open Court, Everyday Mathematics, and the Wright Group)
2. Reed Elsevier (including Harcourt, LexisNexis, Reinhard and Winston, Rigby, Steck-Vaughn, Reading Recovery, Heinemann, and Riverdeep)
3. Vivendi (including Houghton Mifflin, McDougal Littell, Riverside Assessments, Sunburst Technology, and Great Source)[3]
4. Pearson (including Addison Wesley Longman, Scott Foresman, Silver Burdett Ginn, Simon and Schuster, Globe Fearon, Modern Curriculum Press, Celebration Press, Dale Seymour Publications, Prentice Hall School, Waterford Early Reading, Waterford Early Math and Science, Sing, Spell, Read and Write)

We selected two publishers per grade band level (elementary, middle, and high school). Because our independent sources only identified publishers with the largest market share and not specific mathematics curriculum materials, we asked the publishers to select their curricula with the highest market share. The publishers then submitted the curricular materials and accompanying evaluation studies that they had conducted or were aware of for our review.

We analyzed evaluations of the following six commercially generated programs:

Elementary School:

- *Math K-5, 2002* (Addison Wesley/Scott Foresman)
- *Harcourt Math K-6* (Harcourt Brace)

Middle School:

- *Applications and Connections, Grades 6-8, 2001* (Glencoe/McGraw-Hill)
- *An Incremental Development, Sixth Grade, Eighth Grade* (2nd edition) and *An Incremental Development, Seventh Grade, Algebra $^1/_2$ and Algebra* (3rd edition) (Saxon)

---

[3]Houghton Mifflin Company was later sold by Vivendi (in December 2002). Houghton Mifflin Company sold Sunburst Technology in October 2002.

High School:

- *Larson Series, Grades 9-12, 2002* (Houghton-Mifflin/McDougal Littell)
- *University of Chicago School Mathematics Project Integrated Mathematics, Grades 9-12, 2002* (Prentice Hall)

Prentice Hall[4] was an exception and could not choose its curricular materials because we specifically asked for the secondary materials of the University of Chicago School Mathematics Project (UCSMP) to be part of our review. UCSMP was selected because its history and profile represented a "hybrid" between the two categories (NSF-supported and commercially generated curricular programs), and all of its development and research support for the first edition was provided by foundations.[5] We chose UCSMP because, similar to the NSF curricula, its philosophy and program theory are aligned with the National Council of Teachers of Mathematics (NCTM) Standards (NCTM, 1989), although it preceded the NSF-supported curricula in its development period (Thompson et al., 2003). It also differs from the high school NSF-supported materials in that it preserves the traditional course sequence of algebra, geometry, algebra and trigonometry, and advanced mathematics, including newer topics such as statistics and discrete mathematics, whereas the other NSF-supported materials integrate across mathematical subfields at each grade level. UCSMP's development was situated at a university, unlike any other commercially generated curricula. As a result, many published studies and doctoral dissertations were conducted on it.

DATA GATHERING

Information on evaluation studies of the 19 mathematics curricula projects was gathered in several ways. First, we found contacts for all

---

[4]UCSMP, recently acquired by Prentice Hall, received broad NSF support and its secondary program was first headed by Zalman Usiskin and Sharon Senk. It eventually included five components, including an elementary component that produced Everyday Math (headed by Max Bell, with NSF support, and included in this study), a professional development component with NSF support, and an evaluation component headed by Larry Hedges and Susan Stodolsky. In this review, UCSMP refers to the secondary program only, and Everyday Mathematics is coded as EM. Following our charge, UCSMP is categorized as a secondary commercially generated project, whereas EM is categorized as NSF supported. Both had private foundation funding, and for grades 4 through 6 materials, EM received NSF funding.

[5]Amoco Foundation, Carnegie Corporation of New York, and General Electric Foundation.

curricula under review and requested copies of curricular materials and evaluation studies. We received the requested curriculum materials from all publishers except Harcourt Brace. Seventeen of the 19 curricula submitted public evaluation materials to our committee (except *Math K-5, 2002* [Addison Wesley/Scott Foresman] and *Harcourt Math K-6* [Harcourt Brace]). We requested that principal investigators from the NSF-supported mathematics curriculum projects send reports they had submitted to the NSF, as well as their own evaluation studies of their materials or others of which they were aware. We also gathered evaluation studies from all four mathematics NSF Implementation Centers (http://www.ehr.nsf.gov/esie/resources/impsites.asp). We then checked citations and bibliography entries from these studies for possible additional evaluations and acquired copies of new studies. Finally, we conducted library and web searches, and e-mailed several mathematics and mathematics education listservs requesting evaluation studies. We then obtained copies of pertinent studies. A total of 698 evaluation studies were found or submitted for consideration.

We held two evidence-gathering workshops in 2002. The two workshop panels addressed the following questions:

How would you define or evaluate *effectiveness* of a K-5, 6-8, or 9-12 NSF-supported or commercially generated mathematics curriculum?

What evidence would be needed? Be specific in terms of (1) primary and secondary variables, (2) methods of examining or measuring those variables, (3) research designs, and (4) other relationships under investigation.

The first workshop consisted of panels addressing specific topics:

• Evaluation and cross-disciplinary frameworks on curriculum implementation in complex settings;
• Developer, researcher, and evaluator perspectives of curriculum effectiveness;
• The role of content analysis and research on learning in evaluating curricula effectiveness;
• Consideration of professional development needs in curricular implementation; and
• Curriculum decision making and evaluation in school settings.

The second workshop on commercially generated materials asked the same general questions, with two additional requests for comments:

• How do you evaluate materials in relation to the quality and effectiveness of the materials themselves, including content analysis, theories of learning, and teaching? Discuss the role of authors and developers in the process of evaluation.

• How does your company consider the issues of implementation in relation to effectiveness, such as professional development, high-stakes tests, standards, technology, equity, and the adoption of materials and marketing issues?

Much of the debate around curriculum quality in the mathematics and mathematics education community resulted in part because content analysis is an ill-defined concept; therefore, we solicited statements on this topic. Sixteen prominent mathematicians and mathematics educators from a variety of perspectives on content analysis were identified. We sent a written request and received statements from eight: Richard Askey, University of Wisconsin, Madison; Eric Gutstein, University of Illinois, Chicago; Roger Howe, Yale University; William McCallum, University of Arizona; R. James Milgram, Stanford University; Luis Ortiz-Franco, Chapman University; Deborah Schifter, Education Development Center; and Hung Hsi Wu, University of California, Berkeley. We asked for their advice on content analysis by addressing the following questions:

• What should be included in a content analysis?
• How would you judge the quality of a content analysis?
• What is the definition of content analysis?
• Does your response represent the intended and enacted curriculum?
• What references are available in the field on this topic?

THE STUDY MATRIX

We included evaluation studies that focused on one or more of the 13 NSF-supported or 6 commercially generated mathematics curricula, and whose authorship and affiliation were identified. Evaluation studies also had to fall into one of the following categories: (1) Comparative Analysis, (2) Case Study, (3) Content Analysis, (4) Synthesis Study, (5) Background Information, Historical Documentation, or Report to the NSF; and (6) Informative Study (Chapters 4 through 6 provide category descriptions.) We did not wish to limit its initial review to published studies because the topic is relatively current and some papers may not yet have been published. Dissertations would have been excluded if only published studies had been chosen, and we believed these studies could contain useful information. Finally, we sought studies from the following classifications pertaining to curriculum implementation:

- Studies with specific student outcomes
- Content analysis studies
- Studies of classroom implementation and school environment
- Studies of teacher knowledge, teacher characteristics, and professional development

We decided to add to these classifications as we identified additional relevant categories. None were found. The decision tree (Figure 2-1) illustrates the process for categorizing the evaluation studies.

We considered all 698 studies that were found or submitted. If the study met the criteria listed, it was added to the bibliography for review. If it did not meet these criteria, it was placed on a list along with the documented reasons for exclusion. A study whose inclusion was difficult to determine was submitted for committee review. The bibliography of studies that are included in our analysis appears in Appendix B. The 417 studies that met the inclusion criteria for categories 1 through 6 were entered into our database for review.

STUDY CHARACTERISTICS

Table 2-1 shows the distribution of studies by methodology and identifies them as NSF-supported, commercially generated, or UCSMP.

We identified studies that fit the categories of content analysis (n=36), comparative analysis (n=95), case studies (n=45), and synthesis (n=16) as particularly salient. These 192 studies formed the core of the review because they provided direct information pertinent to reviewing the evaluation on materials' effectiveness. Therefore, a large percentage of studies initially considered did not meet the criteria and were excluded from further review. The categories of background information, historical documentation, reports to the NSF, or informative studies were eliminated from further review, though they remain a valuable source of information about program theory and decision making that affected evaluation study designs.

The number of studies in the commercial category was far smaller than the number of studies on the NSF-supported materials or UCSMP.[6] Two factors seem to account for this disparity. First, many NSF- or foundation-supported curricula were required to provide evaluations. The NSF also funded some of these curriculum projects to conduct further evaluation studies. Second, the NSF and UCSMP materials were written primarily by

---

[6]The committee separated UCSMP from the NSF-supported and commercially generated materials because of its hybrid nature.

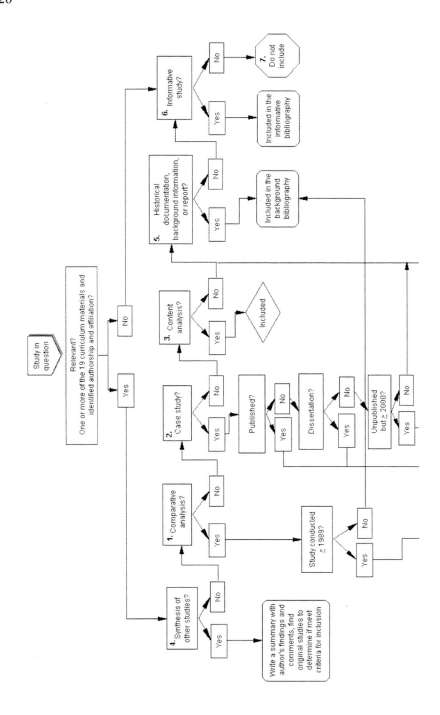

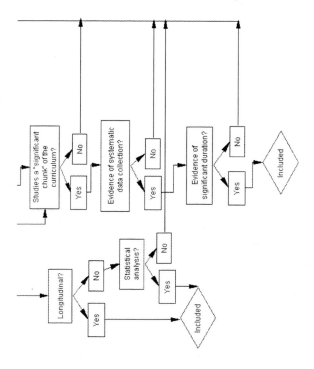

FIGURE 2-1 Decision tree for categorizing evaluation studies.

TABLE 2-1 Distribution of Types of Studies

Type of Study	Number of Studies	Percentage of Total
1. Comparative analysis	95	13.6
NSF-supported curricula	66	69.5
Commercially generated curricula	16	16.8
UCSMP	11	11.6
Not counted in above	2	2.1
2. Case	45	6.4
NSF-supported curricula	45	100.0
Commercially generated curricula	0	0.0
UCSMP	0	0.0
Not counted in above	0	0.0
3. Content analysis	36	5.2
NSF-supported curricula	17	47.2
Commercially generated curricula	1	2.8
UCSMP	12	33.3
Not counted in above	6	16.7
4. Synthesis	16	2.3
NSF-supported curricula	15	93.8
Commercially generated curricula	0	0.0
UCSMP	1	6.3
Not counted in above	0	0.0
5. and 6. Background information and Informative studies	225	32.2
7. Do not include	281	40.3
TOTAL	698	100.0

university faculty whose graduate students often conducted the studies as part of research toward their degrees. Finally, unlike the NSF-supported materials, commercial publishers often conducted market studies, which emphasize how potential purchasers will view the materials. Thus, many commercially generated studies were only marginally useful in evaluating curricular effectiveness.

The evaluation studies were distributed unevenly across the curricula (Table 2-2). Three of the five curricula with the most evaluation studies under review received additional NSF funding to conduct revisions.[7] The elementary, Everyday Mathematics, and secondary components of the UCSMP materials followed.

A database was developed to summarize the studies. Each study was

---

[7]Connected Mathematics, Mathematics In Context, and Core-Plus.

TABLE 2-2 Distribution of Curricula by Study Type

Curriculum Name	Type of Study*					Number of Appearances in Any Study
	1.	2.	3.	4.	5. and 6.	
NSF-supported curricula						
Elementary						
Everyday Mathematics	17	9	7	2	16	51
Investigations in Number, Data and Space	5	1	2	2	9	19
Math Trailblazers	1	1	1	2	6	11
Middle school						
Connected Mathematics Project (CMP)	10	18	8	2	42	80
Mathematics in Context (MiC)	1	8	7	5	52	73
Math Thematics (STEM)	2	6	4	2	13	27
MathScape	0	2	1	1	5	9
Middle School Mathematics Through Applications Project (MMAP)	0	0	0	1	7	8
High school						
Contemporary Mathematics in Context (Core-Plus)	13	5	3	3	19	43
Interactive Mathematics Program (IMP)	17	2	4	2	12	37
Systemic Initiative for Montana Mathematics and Science (SIMMS)	5	1	2	2	10	20
Math Connections	2	0	2	2	6	12
Mathematics: Modeling Our World (MMOW/ARISE)	0	0	2	1	5	8
Commercially generated curricula						
Elementary						
Addison Wesley/Scott Foresman	0	0	2	0	1	3
Harcourt Brace	0	0	1	0	0	1
Middle school						
Saxon	13	0	6	0	21	40
Glencoe/McGraw-Hill	1	0	2	0	4	7
High school						
Prentice Hall/UCSMP	13	0	14	3	46	76
Houghton-Mifflin/ McDougal Littell	2	0	0	0	1	3
Number of evaluation studies	95	45	36	16	225	417
Number of times each curriculum is in each type of study	102	53	68	29	275	528

*Study types: (1.) Comparative Analysis, (2.) Case Study, (3.) Content Analysis, (4.) Synthesis, (5. and 6., respectively) Background Information and Informative Study.

read and analyzed by two or more committee members, the Mathematical
Sciences Education Board staff, or graduate students trained to search for
and record the study characteristics listed in Box 2-1.

After initial review, we studied in depth the first four categories in
Table 2-2 because these studies provided detailed evaluation data.

STUDY CHARACTERISTICS FOR CATEGORIES 1 THROUGH 4

Table 2-3 shows distribution of the published studies by type of study.
Of the studies reviewed, only 22 percent were published in journals or
books. Approximately 28 percent of comparative studies and 31 percent of
case studies were unpublished doctoral dissertations. Although dissertations are unpublished, these studies have been vigorously screened and
provided valuable insight into current evaluation data on the curricula
under review.

TABLE 2-3 Distribution of Published Studies by Type of Study

Type of Study	Number of Studies	Published Study	Unpublished Study	Unpublished Thesis or Dissertation
1. Comparative analysis	95	15	53	27
NSF-supported curricula	66	12	39	15
Commercially generated curricula	16	0	7	9
UCSMP	11	3	6	2
Not counted in above	2	0	1	1
2. Case	45	11	20	14
NSF-supported curricula	45	11	20	14
Commercially generated curricula	0	0	0	0
UCSMP	0	0	0	0
Not counted in above	0	0	0	0
3. Content analysis	36	3	33	0
NSF-supported curricula	17	3	14	0
Commercially generated curricula	1	0	1	0
UCSMP	12	0	12	0
Not counted in above	6	0	6	0
4. Synthesis	16	14	2	0
NSF-supported curricula	15	13	2	0
Commercially generated curricula	0	0	0	0
UCSMP	1	1	0	0
Not counted in above	0	0	0	0
TOTAL	192	43	108	41

BOX 2-1
Study Characteristics

Author(s)

Title and date of publication

Sample program(s) of interest

Comparison curriculum used program

Design of experiment

Author(s) background regarding study

Version of material

Published? Where?

Unit of analysis

Study of duration

Research question

Outcome measures: Student level

Standardized tests

Other measures (attitudes, absentee rates, or dropout rates)

Outcome measures: Teacher level

Content knowledge

Attitude

Student population: Sample and comparison

Total number of students

Gender

Race/ethnicity

Socioeconomic: Free or reduced lunch

Other

Teacher population: Sample and comparison

Total number of teachers

Hours of professional development received

Use of supplemental materials

Mathematics certified

Average number of years of teaching experience

School population: Sample and comparison

Total number of schools

Staff turnover rate

School location (urban, suburban, rural)

Enacted curriculum, measurement, and findings

Author(s) findings/claims

TABLE 2-4 Distribution of Author's Background by Type of Study

Type of Study	Author's Background Regarding Studies				
	Number of Studies	Internal	External	Graduate Student	Unknown
1. Comparative analysis	95	35	27	29	4
NSF-supported curricula	66	20	24	19	3
Commercially generated curricula	16	9	1	6	0
UCSMP	11	5	2	4	0
Not counted in above	2	1	0	0	1
2. Case	45	10	14	14	7
NSF-supported curricula	45	10	14	14	7
Commercially generated curricula	0	0	0	0	0
UCSMP	0	0	0	0	0
Not counted in above	0	0	0	0	0
3. Content analysis	36	16	20	0	0
NSF-supported curricula	17	6	11	0	0
Commercially generated curricula	1	0	1	0	0
UCSMP	12	10	2	0	0
Not counted in above	6	0	6	0	0
4. Synthesis	16	16	0	0	0
NSF-supported curricula	15	15	0	0	0
Commercially generated curricula	0	0	0	0	0
UCSMP	1	1	0	0	0
Not counted in above	0	0	0	0	0
TOTAL	192	77	61	43	11

Table 2-4 shows the distribution of the author's background by type of study. Across the four types of evaluation studies reviewed, 40 percent were done by authors internal to the curriculum project studied. If the study had more than one author, the authors were considered internal if one or more were related to the project (e.g., curriculum developer, curriculum project's evaluator, and curriculum staff). Twenty-two percent of the studies were conducted by graduate students, who may be internal to the project because often they are the curriculum developer's graduate students who perform the research studies. Because the relationship of the graduate student to the curriculum project is not always known, it can only be definitively stated that all authors were external to the project in 32 percent of the studies. The relationship of the author to the curriculum project is unknown in 6 percent of studies.

TABLE 2-5 Distribution of the Study Size by Type of Studies

Type of Study	Number of Studies	Study Size (Students)			
		0-299	300-999	>1,000	n/a
1. Comparative analysis	95	42	28	23	2
2. Case	45	16	3	3	23

Table 2-5 shows the number of students sampled in each of the comparative and case studies (n=140). Studies with fewer than 300 students made up the largest percentage in both comparative and case studies: 44 and 36 percent, respectively. Only 19 percent of the 140 studies had a student sample greater than 1,000 students.

Table 2-6 shows the distribution of sample school locations by type of

TABLE 2-6 Distribution of Sample School Location by Type of Study

Type of Study	Number of Studies	Percent of Studies to Report Data	Percentage of Studies That Reported Location[*]			
			Urban	Suburban	Rural	Only the State/ Region
1. Comparative analysis	95	88	36	46	31	37
NSF-supported curricula	66	89	29	38	24	35
Commercially generated curricula	16	81	31	44	19	38
UCSMP	11	91	40	60	40	20
Not counted in above	2	100	100	50	50	0
2. Case	45	71	27	55	30	30
NSF-supported curricula	45	71	27	55	30	30
Commercially generated curricula	0	0	0	0	0	0
UCSMP	0	0	0	0	0	0
Not counted in above	0	0	0	0	0	0
TOTAL	140	84	33	49	31	35

[*]Many studies report more than one of the following three types of locations: urban, suburban, or rural.

study. Approximately 88 percent of comparative studies and 71 percent of case studies reported data on school location (urban, suburban, rural, or state/region). Suburban students were the largest percentage in both study types. Rural students were the smallest sample in comparative studies, which implies less information is known about curricular effectiveness in these regions. Most studies did not break down the sample by each of the school location types; thus an exact percentage of school types could not be determined. The data that were reported showed wide variation in demographics, although compared with overall representation in the country, minority populations were undersampled (U.S. Department of Education, 2001).

Content analysis studies are not included in Tables 2-5 and 2-6 because they do not report data on students or sample school locations. Synthesis studies are also excluded because they are summaries of multiple studies and typically did not report data on types of schools or students or include data from only some of the studies considered.

Only 19 percent of comparative and case studies provided detailed information on teachers (e.g., certification, years of teaching, or measures of content knowledge) shown in Table 2-7. Generally, comparison groups were based on matching student and not teacher characteristics. Therefore,

TABLE 2-7 Distribution of Studies That Reported Teacher Data by Type of Study

| | | Percent of Reporting Studies | |
Type of Study	Number of Studies	Reported Number of Teachers	Reported Teacher Experience*
1. **Comparative analysis**	**95**	**52**	**14**
NSF-supported curricula	66	56	9
Commercially generated curricula	16	31	19
UCSMP	11	64	36
In 2 categories (not above)	2	0	0
2. **Case**	**45**	**87**	**29**
NSF-supported curricula	45	87	29
Commercially generated curricula	0	0	0
UCSMP	0	0	0
In 2 categories (not above)	0	0	0
TOTAL	140	63	19

*If a study reported on teacher certification or number of years of teaching, it was counted as "Reported Teacher Experience."

some bias may be present in the studies in terms of use of volunteer teachers. A substantial percentage of the studies included some measure of implementation by including teacher logs, classroom observations, interviews, and so forth. However, few included any type of measure of quality of instruction, although case studies were more likely to provide insight into these factors than were comparative studies.

3

Framework for Evaluating Curricular Effectiveness

In this chapter, we present a framework for use in evaluating mathematics curricula. By articulating a framework based on what an effective evaluation could encompass, we provide a means of reviewing the quality of evaluations and identifying their strengths and weaknesses. The framework design was formed by the testimony of participants in the two workshops held by the committee, and by a first reading of numerous examples of studies.

The framework's purpose is also to provide various readers with a consistent and standard frame of reference for defining what is meant by a scientifically valid evaluation study for reviewing mathematics curriculum effectiveness. In addition to providing readers with a means to critically examine the evaluations of curricular materials, the framework should prove useful in guiding the design of future evaluations.

The framework is designed to be comprehensive enough to apply to evaluations from kindergarten through 12th-grade and flexible enough to apply to the different types of curricula included in this review.

With the framework, we established the following description of and definition for curricular effectiveness that is used in the remainder of this report:

> Curricular effectiveness is defined as the extent to which a curricular program and its implementation produce positive and curricularly valid outcomes for students, in relation to multiple measures of students' mathematical proficiency, disaggregated by

content strands and disaggregated by effects on subpopulations of students, and the extent to which these effects can be convincingly or causally attributed to the curricular intervention through evaluation studies using well-conceived research designs. Describing curricular effectiveness involves the identification and description of a curriculum and its programmatic theory and stated objectives; its relationship to local, state, or national standards; subsequent scrutiny of its program contents for comprehensiveness, accuracy and depth, balance, engagement, and timeliness and support for diversity; and an examination of the quality, fidelity, and character of its implementation components.

Effectiveness can be defined in relation to the selected level of aggregation. A single study can examine whether a curricular program is effective (at some level and in some context), using the standards of scientifically established as effective outlined in this report. This would be termed, "a scientifically valid study." Meeting these standards ensures the quality of the study, but a single, well-done study is not sufficient to certify the quality of a program. Conducting a set of studies using the multiple methodologies described in this report would be necessary to determine if a program can be called "scientifically established as effective." Finally, across a set of curricula, one can also discern a similarity of approach, such as a "college preparation approach," "a modeling and applications approach," or a "skills-based, practice-oriented approach," and it is conceivable that one could ask the question of whether an approach is effective, and if so, producing an approach that's "scientifically established as effective." The methodological differences among these levels of aggregation are critical to consider and we address the potential impact of these distinctions in our conclusions.

Efficacy is viewed as considering issues of cost, timeliness and resource availability relative to the measure of effectiveness. Our charge was limited to an examination of effectiveness, thus we did not consider efficacy in any detail in this report.

Our framework merged approaches from method-oriented evaluation (Cook and Campbell, 1979; Boruch, 1997) that focus on issues of internal and external validity, attribution of effects, and generalizability, with approaches from theory-driven evaluations that focus on how these approaches interact with practices (Chen, 1990; Weiss, 1997; Rossi et al., 1999). This permitted us to consider the content issues of particular concern to mathematicians and mathematics educators, the implementation challenges requiring significant changes in practice associated with reform curricula, the role of professional development and teaching capacity, and the need for rigorous and precise measurement and research design.

We chose a framework that requires that evaluations should meet the high standards of scientific research *and* be fully dedicated to serving the information needs of program decision makers (Campbell, 1969; Cronbach, 1982; Rossi et al., 1999). In drawing conclusions on the quality of the corpus of evaluations, we demanded a high level of scientific "validity" and "credibility" because of the importance of this report to national considerations of policy. We further acknowledge other purposes for evaluation, including program improvement, accountability, cost-effectiveness, and public relations, but do not address these purposes within our defined scope of work. Furthermore, we recognize that at the local level, decisions are often made by weighing the "best available evidence" and considering the likelihood of producing positive outcomes in the particulars of context, time pressures, economic feasibility, and resources. For such purposes, some of the reported studies may be of sufficient applicability. Later in this section, we discuss these issues of utility and feasibility further and suggest ways to maintain adequate scientific quality while addressing them.

Before discussing the framework, we define the terms used in the study. There is ambiguity in the use of the term "curriculum" in the field (National Research Council [NRC], 1999a). In many school systems, "curriculum" is used to refer to a set of state or district standards that broadly outline expectations for the mathematical content topics to be covered at each grade level. In contrast, at the classroom level, teachers may select curricular programs and materials from a variety of sources that address these topics and call the result the curriculum. When a publisher or a government organization supports the development of a set of materials, they often use the term "curriculum" to refer to the physical set of materials developed across grade levels. Finally, the mathematics education community often finds it useful to distinguish among the intended curriculum, the enacted curriculum, and the achieved curriculum (McKnight et al., 1987). Furthermore, in the curriculum evaluation literature, some authors take the curriculum to be the physical materials and others take it to be the physical materials together with the professional development needed to teach the materials in the manner in which the author intended. Thus "curriculum" is used in multiple ways by different audiences.

We use the term "curriculum" or "curricular materials" in this report as follows:

> A curriculum consists of a set of materials for use at each grade level, a set of teacher guides, and accompanying classroom assessments. It may include a listing of prescribed or preferred classroom manipulatives or technologies, materials for parents, homework booklets, and so forth. The curricula reviewed in this report are

written by a single author or set of authors and published by a single publisher. They usually include a listing or mapping of the curricular objectives addressed in the materials in relation to national, state, or local standards or curricular frameworks.

We also considered the meaning of an evaluation of a curriculum for the purposes of this study. To be considered an evaluation, a curriculum evaluation study had to:

- Focus primarily on one of the curriculum programs or compare two or more curriculum programs under review;
- Use a methodology recognized by the fields of mathematics education, mathematics, or evaluation; and
- Study a major portion of the curriculum program under investigation.

A "major portion" was defined as at least one grade-level set of materials for studies of intended curricular programs and a significant piece (more than one unit) of curricular materials and a significant time duration of use (at least a semester) for studies of enacted curricular programs. Evaluation studies were also identified and distinguished from research studies by requiring evaluation studies to include statements about the effectiveness of the curriculum or suggestions for revisions and improvements. Further criteria for inclusion or exclusion were developed for each of the four classes of evaluation studies identified: content analyses, comparative analyses, case studies, and synthesis studies. These are described in detail in Chapters 4 through 6. Many formative, as opposed to summative, assessments were not included.

The framework we proposed consists of two parts: (1) the components of curricular evaluation (Figure 3-1), and (2) evaluation design, measurement, and evidence, (Figure 3-2). The first part guides an evaluator in specifying the program under investigation, while the second part articulates the methodological design and measurement issues required to ensure adequate quality of evidence. Each of these two parts is described in more detail in this chapter.

The first part of the framework consists of primary and secondary components. The primary components are presented in Figure 3-1: program components, implementation components, and student outcomes. Secondary components of the framework include systemic factors, intervention strategies, and unanticipated influences.

The second part of the framework, evaluation design, measurement, and evidence, is divided into articulation of program theory, selection of research design and methodology, and other considerations.

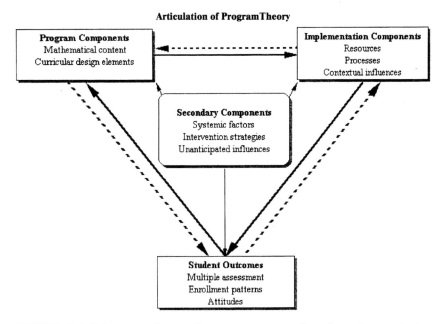

FIGURE 3-1 Primary and secondary components of mathematics curriculum evaluation.

PRIMARY COMPONENTS

For each of the three major components (program, implementation, and student outcomes), we articulated a set of subtopics likely to need consideration.

Program Components

Examining the evaluation studies for their treatments of design elements was a way to consider explicitly the importance, quality, and sequencing of the mathematics content. Our first consideration was the major theoretical premises that differentiate among curricula. Variations among the evaluated curricula include the emphasis on context and modeling activities: the importance of data; the type and extent of explanations given; the role of technology; the importance of multiple representations and problem solving; the use and emphasis on deductive reasoning, inductive reasoning, conjecture, refutation, and proof; the relationships among the mathematical subfields such as algebra, geometry, and probability; and the focus on calculation, symbolic manipulations, and conceptual development. Views of learning and teaching, the role of practice, and the directness of

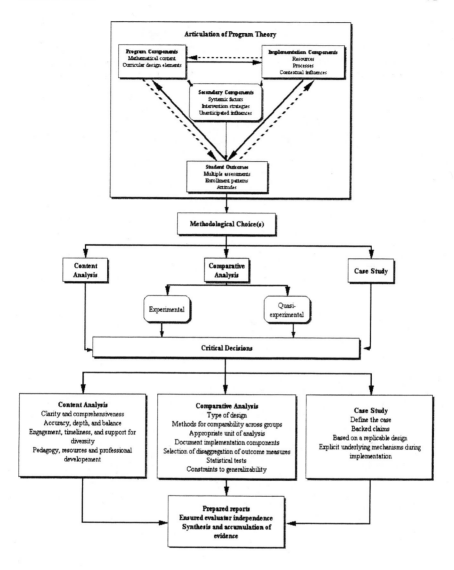

FIGURE 3-2 Framework for evaluating curricular effectiveness.

instruction also vary among programs. It is important for evaluators to determine these differences and to design evaluations to assess the advantages and disadvantages of these decisions in relation to student learning.

At the heart of evaluating the quality of mathematics curriculum materials is the analysis of the mathematical content that makes up these materials. We call this "content analysis" (Box 3-1). A critical area of debate in conducting a content analysis is how to assess the trade-offs among the various choices. Curricular programs must be carried out within the constraints of academic calendars and school resources, so decisions on priorities in curricular designs have real implications for what is subsequently taught in classrooms. An analysis of content should be clear and specific as to what trade-offs are made in curricular designs.

A second source of controversy evolves from a debate over the value of conducting content analysis in isolation from practice. Some claim that until one sees a topic taught, it is not really possible to specify what is

---

BOX 3-1
Factors to Consider in Content Analysis of
Mathematics Curriculum Materials

Listing of topics

Sequence of topics

Clarity, accuracy, and appropriateness of topic presentation

Frequency, duration, pace, depth, and emphasis of topics

Grade level of introduction

Overall structure: integrated, interdisciplinary, or sequential

Types of tasks and activities, purposes, and level of engagement

Use of prior knowledge, attention to (mis)conceptions, and student strategies

Reading level

Focus on conceptual ideas and algorithmic fluency

Emphasis on analytic/symbolic, visual, or numeric approaches

Types and levels of reasoning, communication, and reflection

Type and use of explanation

Form of practice

Approach to formalization

Use of contextual problems and/or elements of quantitative literacy

Use of technology or manipulatives

Ways to respond to individual differences and grouping practices

Formats of materials

Types of assessment and relation to classroom practice

learned (as argued by William McCallum, University of Arizona, and Richard Lehrer, Vanderbilt University, when they testified to the committee on September 18, 2002). In this sense, a content analysis would need to include an assessment of what a set of curricular tasks makes possible to occur in a classroom as a result of activity undertaken, and would depend heavily on the ability of the teacher to make effective use of these opportunities and to work flexibly with the curricular choices. This kind of content analysis is often a part of pilot testing or design experiments. Others prefer an approach to content analysis that is independent of pedagogy to ensure comprehensiveness, completeness, and accuracy of topic and to consider if the sequencing forms a coherent, logical, and age-appropriate progression. Both options provide valuable and useful information in the analysis of curricular effectiveness but demand very different methodologies.

Another consideration might be the qualifications of the authors and their experience with school and collegiate mathematics. The final design element concerns the primary audience for curricular dissemination. One publisher indicated its staff would often make decisions on curricular design based on the expressed needs or preferences of state adoption boards, groups of teachers, or in the case of home schooling, parents. Alternatively, a curriculum might be designed to appeal to a particular subgroup, such as gifted and talented students, or focus on preparation for different subsequent courses, such as physics or chemistry.

Implementation Components

Curricular programs are enacted in a variety of school settings. Curriculum designers consider these settings to various degrees and in various ways. For example, implementation depends heavily on the capacity of a school system to support and sustain the curriculum being adopted. This implies that a curricular program's effectiveness depends in part on if it is implemented adequately and how it fits within the grade-level band for which it is designed as well as whether it fits with the educational contexts that proceed or follow it.

Implementation studies have provided highly convincing evidence that implementation is complicated and difficult because curricula are enacted within varying social contexts. Factors such as participation in decision making, incentives such as funding or salaries, time availability for professional development, staff turnover or student attendance, interorganizational arrangements, and political processes can easily hinder or enhance implementation (Chen, 1990).

In evaluation studies, these issues are also referred to as process evaluation or program or performance monitoring. Implementation includes examining the congruity between the instruction to students and the goals of the program, whether the implemented curriculum is reaching all students, how well the system is organized and managed to deliver the curricular program, and the adequacy of the resources and support. Process evaluation and program and performance monitoring are elements of program evaluation that can provide essential data in judging the effectiveness of the program and in providing essential feedback to practitioners on program improvement (Rossi et al., 1999).

In the use of curricula in practice, many variations enter the process. We organized the factors in the implementation component into three categories: resource variables, process variables, and community/cultural influences. Examples of each are listed in Table 3-1.

Resource variables refer to the resources made available to assist in implementation. Process variables refer to the ways and means in which implementation activities are carried out, decisions are made, and information is analyzed on the practice and outcomes of teaching mathematics. Community and cultural factors refer to the social conditions, beliefs, and expectations held both implicitly and explicitly by participants at the site of adoption concerning learning, teaching, and assessing student work and opportunities.

We also identified a set of mediating factors that would be most likely to influence directly the quality and type of implementation.

Appropriate Assignment of Students

Decisions concerning student placement in courses often have strong implications for the success of implementation efforts and the distribution of effects across various student groups. Evaluations must carefully document and monitor the range of student preparation levels that teachers must serve, the advice and guidance provided to students and parents as to what curricular choices are offered, and the levels of attrition or growth of student learning experienced over a curricular evaluation study period by students or student subpopulations.

Ensuring Adequate Professional Capacity

This was viewed as so critical to the success of implementation efforts that some argued that its significance exceeds that of curriculum in determining students' outcomes (as stated by Roger Howe, Yale University, in testimony to the committee at the September 2002 workshop). Professional capacity has a number of dimensions. First, it includes the number and qualifications of the actual teachers who will instruct students. Many new curricula rely on teachers' knowledge of topics that were not part of their own education. Such topics could include probability and statistics, the use

TABLE 3-1 Categories and Examples of Implementation Component Variables

Resource Variables	Process Variables	Community/Cultural Influences
Teacher supply, qualifications, and rate of turnover	Teacher organization and professional community	Teacher beliefs concerning learning, teaching, and assessment
Professional development and teacher knowledge	Curricular decision making	Expectations of schooling and future educational and career aspirations
Length of class	Course requirements	Homework time
Class size and number of hours of preparation per teacher	Course placements, guidance, and scheduling	Stability, language proficiency, and mobility of student populations
Cost and access to materials, manipulatives, and technology	Administrative or governance of school decision making	Combinations of ethnic, racial, or socioeconomic status among students, teachers, and community
Frequency and type of formative and summative assessment practices	Forms and frequency of assessment and use of data	Community interest and responses to publicly announced results on school performance
Extent and type of student needs and support services		Student beliefs and expectations
Parental involvement		Parental beliefs and expectations

of new technologies, taking a function-based approach to algebra, using dynamic software in teaching geometry, contextual problems, and group methods. In addition, school districts are facing increasing shortages of mathematics teachers, so teachers frequently are uncertified or lack a major in mathematics or a mathematics-related field (National Center for Education Statistics [NCES], 2003). At the elementary level, many teachers are assigned to teach all subjects, and among those, many are required to teach mathematics with only minimal training and have a lack of confidence or affection for the discipline (Ma, 1999; Stigler and Hiebert 1999). Finally, especially in urban and rural schools, there is a high rate of teacher turnover

(Ingersoll, 2003; National Commission on Teaching and America's Future, 2003), so demanding curricula may not be taught as intended or with consistency over the full duration of treatment.

Also in this category are questions of adequate planning time for implementing and assessing new curricula and adequate support structures for trying new approaches, including assistance, reasonable class sizes, and number of preparations. Furthermore, if teachers are not accorded the professional opportunities to participate in decision making on curricular choices, resistance from them, reverting to the use of supplemental materials with which teachers are more comfortable, or lack of effort can hamper treatment consistency, duration, and quality. In contrast, in some cases, reports were made of teacher-initiated and -dominated curricular reform efforts where the selection, adaptation, and use of the materials was highly orchestrated and professionally evaluated by practitioners, and their use of the materials typically was reported as far more successful and sustainable (as stated by Terri Dahl, Charles M. Russell High School, MT, and Timothy Wierenga, Naperville Community Unit School District #203, IL, in testimony to the committee on September 18, 2002).

Opportunities for professional development also vary. Documenting the duration, timing, and type of professional development needed and implemented is essential in the process of examining the effectiveness of curricular programs. Because many of the programs require that teachers develop new understandings, there is a need for adequate amounts of professional development prior to implementation, continued support during implementation, and reflective time both during and after implementation. Because many of these programs affect students from multiple grade levels, there is also the issue of staging, to permit students to enter the program and remain in it and to allow teachers at higher grade levels to know that students have the necessary prerequisites for their courses.

Finally, there are different types of professional development with varying amounts of attention to content, pedagogy, and assessment (Loucks-Horsley et al., 1998). These involve different amounts of content review and use of activities. In some, the teachers are shown the materials and work through a sample lesson, concentrating on management and pedagogy. In others, teachers work through all units and the focus is on their learning of the content. In a few, the teachers are taught the immediate content and provided coursework to ensure they have learned more of the content than is directly located in the materials (NCES, 1996). If limited time and resources devoted to professional development make the deeper treatments of content infrequent, then this can limit a teacher's capacity to use new materials.

Opportunity to Learn

Not all curricula materials are implemented in the same way. In some schools, the full curricula are used, while in others units may be skipped or time limitations at the end of the year may necessitate abandoning the last units in the sequence. Even within units, topics can be dropped at teachers' discretion. Thus, it is important for evaluations of curricula to document what teachers teach (Porter, 1995). Opportunity to learn is a particularly important component of implementation because it is critically involved in examining differential student performance on particular outcomes. It can be evaluated directly using classroom observation or indirectly through teacher and student logs, or by surveying teachers and students on which items on an end-of-year test were covered.

Instructional Quality and Type

It is also necessary to examine how well curricula are taught. As noted by Stigler and Hiebert (1999, pp. 10-11), "What we can see clearly is that American mathematics teaching is extremely limited, focused for the most part on a very narrow band of procedural skills. Whether students are in rows working individually or sitting in groups, whether they have access to the latest technology or are working only with paper and pencil, they spend most of their time acquiring isolated skills through repeated practice."

Using different curricula may contribute to the opportunity to teach differently, but it is unlikely to be sufficient to change teaching by the mere presence of innovative materials. Typically, teachers must learn to change their practices to fit the curricular demands. In addition to materials, they need professional development, school-based opportunities to plan and reflect on their practices, and participation in professional societies as sources of additional information and support. There is considerable variation in teaching practices, and while one teacher may shift his/her practice radically in response to the implementation of a new curriculum, many will change only modestly, and a substantial number will not alter their instructional practices at all. Evaluation studies that consider this variable typically include classroom observation.

Assessment

Formative or embedded assessment refers to the system of classroom assessment that occurs during the course of normal classroom teaching and is designed in part to inform subsequent instruction. Often assessments are included in instructional materials, and attention must be paid to how these are used and what evidence they provide on curricular effectiveness. It is particularly helpful when teachers use multiple forms of assessment, so that one can gauge student progress across the year on tests, quizzes, and projects, including conceptual development, procedural fluency, and applications. It is also important that such testing provides evidence on the variation among students in learning and creates a longitudinal progression on individual student learning. More and more, schools and school districts are working to coordinate formative assessment with high-stakes testing programs, and when this is done, one can gain an important window on the relationship between curricula and one standardized form of student outcomes.

Because it is in schools with large numbers of students performing below expected achievement levels that the high-stakes testing and accountability models exert the most pressure, it is incumbent upon curriculum evaluators to pay special attention to these settings. It has been widely documented that in urban and rural schools with high levels of poverty, students are likely to be given inordinate amounts of test preparation, and are subject to pull-out programs and extra instruction, which can detract from the time devoted to regular curricular activities (McNeil and Valenzuela, 2001). This is especially true for schools that have been identified as low performing and in which improving scores on tests is essential to ensure that teachers and administrators do not lose their jobs.

Parental Influence and Special Interest Groups

Parents and other members of the community influence practices in ways that can significantly and regularly affect curriculum implementation. The influence exerted by parents and special interest groups differs from systemic factors in that they are closely affiliated with the local school, and can exert pressure on both students and school practitioners.

Parents are influential and need to be considered in multiple ways. Parents provide guidance to students in course selection, they convey differing levels of expectation of performance, and they provide many of the supplemental materials, resources (e.g., computer access at home), and opportunities for additional education (informal or nonschool enrollments). In some cases, they directly provide home schooling or purchase schooling (distance education, Scholastic Aptitude Test [SAT] prep courses) for their children. It is also important to recognize that there are significant cultural differences among parents in the degree to which they will accept or challenge curricular decisions based on their own views of schooling, authority, and sense of welcome within the school or at public meetings (Fine, 1993). To ensure that parental satisfaction and concerns are adequately and fairly considered, evaluators must provide representative sources of evidence.

Measures of Student Outcomes

In examining the effectiveness of curricula, student outcome measures are of critical importance. One must keep in mind that the outcomes of curricular use should be the documentation of the growth of mathematical thinking and knowledge over time. We sought to identify the primary factors that would influence the perceived effectiveness of curricula based on those measures.

Most curricula evaluators use tests as the primary tool for measuring curricular effectiveness. Commonly used tests are large-scale assessments from state accountability systems; national standardized tests such as the SAT, Iowa Test of Basic Skills (ITBS), or AP exams or the National Assessment of Educational Progress (NAEP); or international tests such as items from the Third International Mathematics and Science Study (TIMSS). Among these tests, there are different choices in terms of selecting norm-referenced measures that produce national percentile rankings or criteria-referenced measures that are also subject to scaling influences. At a more specific level, assessments are used that measure particular cognitive skills such as problem solving or computational fluency or are designed to elicit likely errors or misconceptions. At other times, evaluators report outcomes using the tests and assessments included with the curriculum's materials, or develop their own tests to measure program components.

For evaluation studies, there can be significant problems associated with the selection and use of outcome measures. We use the term "curricular validity of measures" to refer to the use of outcome measures sensitive to the curriculum's stated goals and objectives. We believe that effectiveness must be judged relative to curricular validity of measures as a standard of scientific rigor.[1] In contrast, local decision makers may wish to gauge how well a curricular program supports positive outcomes on measures that facilitate student progress in the system, such as state tests, college entrance exams, and future course taking. We refer to this as "curricular alignment with systemic factors."[2]

One should not conflate curricular validity of measures with curricular alignment with systemic factors in evaluation studies. Additionally, whereas the use of measures demonstrating curricular validity is obligatory to determine effectiveness, curricular alignment with systemic factors may also be advised.

---

[1]Thompson et al. (2003) is one example for addressing this issue.

[2]Sconiers et al. (2002) is one example of how results on outcomes of state tests are interpreted in relation to local contexts.

BOX 3-2
Dimensions of Assessments to Be Considered in
Measuring Curricular Effectiveness

Curricular topics

Question format: open ended, structured response, multiple choice

Strands of mathematical proficiency*

Cognitive level

Cognitive development over time

Reliance on prior knowledge or reading proficiency

Length and timing of testing

Familiarity of task

Opportunity to learn

---

*We are referring to "mathematical proficiency" as defined in *Adding It Up: Helping Children Learn Mathematics* (NRC, 2001a).

An evaluator also needs to consider the possibility of including multiple measures; collecting measures of prior knowledge from school, district, and state databases; and identifying methods of reporting the data. To address this variety of measures, we produced a list of dimensions of assessments that can be considered in selecting student outcome measures (Box 3-2). In analyzing evaluations of curricular effectiveness, we were particularly interested in measures that could produce disaggregation of results at the level of common content strands because this is the most likely means of providing specific information on student success on certain curricular objectives. We also scrutinized evaluations of curricula for the construct validity, reliability, and fairness of the measures they employed. For measures of open-ended tasks, we looked for measures of interrater reliability and the development of a clear and precise rubric for analysis. Also important to gauge effectiveness would be longitudinal measures, a limited resource in most studies.

Another critical issue is how to present results. Results may be presented as mean scores with standard deviations, percentage of students who pass at various proficiency or performance levels, or gain scores. We also considered in our analysis of curriculum evaluations whether the results presented in those studies were accompanied by clear specification of the purpose or purposes of a test and how the test results were used in grading or accountability systems. Finally, we considered whether evaluations in-cluded indications of attrition through the course of the study and explana-tions of how losses might influence the data reports.

After considering the selection of indicators of student learning, we explored the question of how those indicators present that information across diverse populations of students. Disaggregation of the data by sub-groups—including gender, race, ethnicity, economic indicators, academic performance level, English-language learners, and students with special needs—can ensure that measures of effectiveness include considerations of equity and fairness. In considering issues of equity, we examined whether evaluations included comparisons between groups—by subgroup on gain scores—to determine the distribution of effects. We also determined whether evaluations reported on comparisons in gains or losses among the subpopu-lations of any particular treatment, to provide evidence on the magnitude of the achievement gap among student groups. Accordingly, we asked whether evaluations examined distributions of scores rather than simply attending to the percentage passing or "mean performance." Did they consider the performance of students at all levels of achievement?

Conducting evaluations of curricula in schools or districts with high levels of student mobility presents another challenge. We explored whether a pre- or postevaluation design could be used to ensure actual measurement of student achievement in this kind of environment. If longitudinal studies were conducted, were the original treatment populations maintained over time, a particularly important concern in schools where there is high mobil-ity, choice, or dropout problems?

Finally, in drawing conclusions about effectiveness, we considered whether the evaluations of curricula employed the use of multiple types of measures of student performance other than test scores (e.g., grades, course-taking patterns, attitudes and perceptions, performance on subsequent courses and postsecondary institutions, and involvement in extracurricular activities relevant to mathematics). An effective curriculum should make it feasible and attractive to pursue future study in a field and to recover from prior failure or enter or advance into new opportunities.

We also recognized the potential for "corruptibility of indicators" (Rossi et al., 1999). This refers to the tendency for those whose perfor-mance is monitored to improve the indicator in any way possible. When some student outcome measures are placed in an accountability system, especially one where the students' retention or denial of a diploma is at stake, the pressure to teach directly to what is likely to be assessed on the tests is high. If teachers and administrators are subject to loss of employ-ment, the pressures increase even more (NRC, 1999b; Orfield and Kornhaber, 2001).

SECONDARY COMPONENTS

Systemic Factors

Systemic factors refer to influences on curricular programs that lie outside the intended and enacted curricula and are not readily amenable to change by the curriculum designers. Examples include the standards and the accountability system in place during the implementation period. Standards are designed to provide a governance framework of curricular objectives in which any particular program is embedded. Various sets of standards (i.e., those written by National Council of Teachers of Mathematics [NCTM], those established by states or districts) represent judgments about what is important. These can differ substantially both in emphasis and in specificity.

Accountability systems are a means to use the combination of standards and typically high-stakes test to conform in offerings across a governance structure. As mentioned previously, curriculum programs may vary in relation to their alignment to standards and accountability systems.

Other policies such as certification and recertification processes or means of obtaining alternative certification for teachers can have a major impact on a curricular change, but these are typically not directly a part of the curriculum design and implementation processes. State requirements for course taking, college entrance requirements, college credits granted for Advanced Placement (AP) courses, ways of effecting grade point averages (i.e., an additional point toward the Grade Point Average [GPA] for taking an AP course), availability of SAT preparation courses, and opportunities for supplementary study outside of school are also examples of systemic factors that lie outside the scope of the specific program, but exert significant influences on it. Textbook adoption processes and cycles also would be considered systemic factors. As reported to the committee by publishers, the timing and requirements of these adoption processes, especially in large states such as Texas, California, and Florida, exert significant pressure on the scope, sequence, timing, and orientation of the curricula development process and may include restrictions on field testing or dates of publication (as reported by Frank Wang, Saxon Publishers Inc., in testimony to the committee at the September 2002 workshop).

Intervention Strategies

Intervention strategies refer to the theory of action that lies behind the enactment of a curricular innovation. We found it necessary to consider attention given by evaluation studies to the intervention strategies behind many of the programs reviewed. Examining an intervention strategy requires one to consider the intent, the strategy, and the purpose and extent of impact intended by the funders or investors in a curricular program.

A program of curricular development supported by a governmental agency like the National Science Foundation (NSF) may have various goals, which can have a significant impact on the conclusions about their effectiveness when undertaking a meta-analysis of different curricula. A curricular development program could be initiated with the goal to improve instruction in mathematics across the board, and thus demand that the developers be able to demonstrate a positive impact on most students. Such an approach views the intent of the curricular initiative as comprehensive school improvement, and one would therefore examine the extent of use of the programs and the overall performance of students subjected to the curriculum over the time period of the intervention.

Alternatively, one could conceive of a program's goal as a catalytic demonstration project, or proof of concept, to establish a small but novel variation to typical practice. In this case, evaluators might not seek broad systemic impact, but concentrate on data supplied by the adopters only, and perhaps only those adopters who embraced the fullest use of the program across a variety of school settings. Catalytic programs might concentrate on documenting a curriculum's potential and then use dissemination strategies to create further interest in the approach.[3]

A third strategy often considered in commercially generated curriculum materials might be to gain market share. The viability of a publishing company's participation in the textbook market depends on the success it has in gaining and retaining sufficient market share. In this case, effectiveness on student performance may be viewed as instrumental to obtaining the desired outcome, but other factors also may be influential, such as persuading decision makers to include the materials in the state-approved list, cost-effectiveness, and timing. Judgments on intervention strategies are likely to have a significant influence on the design and conduct of an evaluation.

Unanticipated Influences

Another component of the framework is referred to as unanticipated influences. During the course of implementation of a curriculum, which may occur over a number of years, factors may emerge that exert significant forces on implementation. For example, if calculators suddenly are permit-

---

[3]This view of a catalytic program was stated by Mark St. John, Inverness Research Associates, in testimony to the committee in September 2002.

ted during test taking, this change may exert a sudden and unanticipated influence on the outcome measures of curricula effectiveness. Equally, if an innovation such as block scheduling is introduced, certain kinds of laboratory-based activities may become increasingly feasible to implement. A third example is the use of the Internet to provide parents and school board members with information and positions on the use of particular materials, an approach that would not have possible a decade ago.

In Figure 3-2, an arrow links student outcomes to other components to indicate the importance of feedback, interactions, and iterations in the process of curricular implementation. Time elements are important in the conduct of an evaluation in a variety of ways. First of all, curricular effects accrue over significant time periods, not just months, but across academic years. In addition, the development of materials undergoes a variety of stages, from draft form to pilot form to multiple versions over a period of years. Also, developers use various means of obtaining user feedback to make corrections and to revise and improve materials.

EVALUATION DESIGN, MEASUREMENT, AND EVIDENCE

After delineating the primary and secondary components of the curriculum evaluation framework, we focused on decisions concerning evaluation and evidence gathering. We identified three elements of that process: articulation of program theory, selection of research design and methodology, and other considerations. These included independence of evaluators, time elements, and accumulation of knowledge and the meta-analysis.

Articulation of Program Theory

An evaluator must specify and clearly articulate the evaluation questions and elaborate precisely what elements of the primary and secondary components will be considered directly in the evaluation, and these elaborations can be referred to as specifying "the program theory" for the evaluation. As stated by Rossi et al. (1999, p. 102):

> Depiction of the program's impact theory has considerable power as a framework for analyzing a program and generating significant evaluation questions. First the process of making that theory explicit brings a sharp focus to the nature, range, and sequence of program outcomes that are reasonable to expect and may be appropriate for the evaluator to investigate.

According to Weiss (1997, p. 46), "programmatic theory . . . deals with mechanisms that intervene between the delivery of the program service and the occurrence of outcomes of interest." Thus, program theory specifies the

---

BOX 3-3
Focus Topics from the 1989 NSF Request for Proposals

New mathematics topics

Links between mathematics and other disciplines

Increased access for underserved students and elimination of tracking

Use of student-centered pedagogies

Increased uses of technologies

Application of research on student learning

Use of open-ended assessments

SOURCE: NSF (1989).

---

evaluator's view of the causal links and covariants among the program components. In terms of our framework, program theory requires the precise specification of relationships among the primary components (program components, implementation components, and student outcomes) and the secondary components (systemic factors, intervention strategies, and unanticipated influences).

For example, within the NSF-supported curricula, there were a number of innovative elements of program theory specified by the Request for Proposals (Box 3-3). For example, the call for proposals for the middle grades curricula specified that prospective developers consider curriculum structure, teaching methods, support for teachers, methods and materials for assessment, and experiences in implementing new materials (NSF, 1989).

In contrast, according to Frank Wang of Saxon Publishing (personal communication, September 11, 2003), their curriculum development and production efforts follow a very different path. The Saxon product development model is to find something that is already "working" (meaning curriculum use increases test scores) and refine it and package it for wider distribution. They see this as creating a direct product of the classroom experience rather than designing a program that meets a prespecified set of requirements. Also, they prefer to select programs written by single authors rather than by a team of authors, which is more prevalent now among the big publishers.

The Saxon pedagogical approach includes the following:

- Focus on the mastery of basic concepts and skills
- Incremental development of concepts[4]
- Continual practice and review[5]
- Frequent, cumulative testing[6]

The Saxon approach, according to Wang, is a relatively rigid one of daily lesson, daily homework, and weekly cumulative tests. At the primary grades, the lesson is in the form of a scripted lesson that a teacher reads; Saxon believes the disciplined structure of its programs is the source of their success.

In devising an evaluation of each of these disparate programs, an evaluator would need to concentrate on different design principles and thus presumably would have a different view of how to articulate the program's theory, that is, why it works. In the first case, particular attention might be paid to the contributions of students in class and to their methods and strategies of working with embedded assessments; in the second case, more attention would be paid to the varied paces of student progress and to the way in which the student demonstrated mastery of both previous and current topics. The program theory would be not simply a delineation of the philosophy and approach of the curriculum developer, but the way in which the measurement and design approach carefully considered those design elements.

Chen (1990) argues that by including careful development of program theory, one can increase the trustworthiness and generalizability of the evaluation study. Trustworthiness, or the view that the results will provide convincing evidence to stakeholders, is increased because with explicit monitoring of program components and interrelationships, evaluators can examine whether outcomes are sensitive to changes in interventions and process variables with greater certainty. Generalizability, or application of results to future pertinent circumstances, is increased because evaluators can determine the extent to which a new situation approximates the one in which the prior result was obtained. Only by clearly articulating program theory and then testing competing hypotheses can evaluators disentangle these complex issues and help decision makers select curricula on the basis of informed judgment.

---

[4]Larger concepts are broken down into small subconcepts that are covered in individual daily lessons that are spread out throughout the year.

[5]Concepts are reviewed and practiced in the daily homework, called problem sets.

[6]Tests are weekly and each is cumulative so that each test is a mini final exam.

Selection of Research Design and Methodology

As indicated in *Scientific Research in Education* (NRC, 2002), scientific evaluation research on curricular effectiveness can be conducted using a variety of methodologies. We focused on three primary types of evaluation design: content analyses, comparative studies, and case studies. (A fourth type, synthesis studies, is discussed under "Accumulation of Knowledge and the Meta-Analysis," later in this chapter.) Typically, content analyses concentrate on program components while case studies tend to elaborate on issues connected to implementation. Comparative studies involve all three major components—program, implementation, and student outcomes—and tend to link them to compare their effects. Subsequent to describing each, a final section on syntheses studies, modeling, and meta-analysis is also provided to complete the framework. Our decision to focus on these three methodologies should not be understood to imply the rejection of other possibilities for investigating effectiveness, but rather a discussion of the most common forms submitted for review. Also, we should note that some evaluations incorporate multiple methodologies, often designing a comparative study of a limited number of variables and supplementing it with the kinds of detailed information found in the case studies.

Content Analyses

Evaluations that focus almost exclusively on examining the content of the materials were labeled content analyses. Many of these evaluations were of the type known as connoisseurial assessments because they relied nearly exclusively on the expertise of the reviewer and often lacked an articulation of a general method for conducting the analysis (Eisner, 2001). Generally, evaluators in these studies reviewed a specific curriculum for accuracy and for logical sequencing of topics relative to the expert knowledge. Some evaluators explicitly contrasted the curriculum being analyzed to international curricula in countries in which students showed high performance on international tests. In our discussions of content analysis in Chapter 4, we specify a number of key dimensions, while acknowledging that as connoisseurial assessments, they involve judgment and values and hence depend on one's assessment of the qualifications and reputation of the reviewer. By linking these to careful examination of empirical studies of the classroom, one can test some of these assumptions directly.

Comparative Studies

A second approach to evaluation of curricula has been for researchers to select pertinent variables that permit a *comparative study* of two or more curricula and their effects over significant time periods. In this case, investigators typically have selected a relatively small number of salient variables from the framework for their specific program theory and have designed or identified tools to measure these variables. The selection of variables was often critical in determining if a comparative study was able to provide explanatory information to accompany its conjectures about causal inference. Many of the subsequent sections of this chapter apply directly to comparative study, but can inform the selection, conduct, and review of case studies or content analyses.

Our discussion of comparative studies focuses on seven critical decisions faced by evaluators in the conduct of comparative studies: (1) select the study type: quasi-experimental or experimental, (2) establish comparability across groups, (3) select a comparative unit of analysis, (4) measure and document implementation fidelity, (5) conduct an impact assessment and choice of outcome measures, (6) the select and conduct statistical tests, and (7) determine limitations to generalizability in relation to sample selection. After identifying the type of study, the next five decisions relate to issues of internal validity, while the last one focuses on external validity. After introducing an array of comparative designs, each of these is discussed in relation to our evaluation framework.

Comparative Designs

In comparative studies, multiple research designs can be utilized, including:

- *Randomized field trials.* In this approach, students or other units of analysis (e.g., classrooms, teachers, schools) are randomly assigned to an experimental group, to which the intervention is administered, and a control group, from which the intervention is withheld.
- *Matched comparison groups.* In this approach, students who have been enrolled in a curricular program are matched on selected characteristics with individuals who do not receive the intervention to construct an "equivalent" group that serves as a control.
- *Statistically equated control.* Participants and nonparticipants, not randomly assigned, are compared, with the difference between them on selected characteristics adjusted by statistical means.
- *Longitudinal studies.* Participants who receive the interventions are compared before and after the intervention and possibly at regular intervals during the treatment.
- *Generic controls.* Intervention effects are compared with established norms about typical changes in the target populations using indicators that are widely available.

The first type of study is referred to as "randomized experiments" and the other four are viewed as "quasi-experiments" (Boruch, 1997). The experimental approach assumes that all extraneous confounding variables will be equalized by the process of random assignment. As recognized by Cook (in press), randomized field trials will only produce interpretable results if one has a clear and precise description of the program, and one can ensure the fidelity of the treatment for the duration of the experiment. Developing practical methods that can ensure these conditions and thus make use of the power of this approach could yield more definitive causal results, especially if linked to complementary methodologies aiding in explanatory power.

Threats to validity in the quasi-experimental approaches are that the selection of relevant variables for comparison may not consider differences that actually affect the outcomes systematically (Agodini and Dynarski, 2001). For example, in many matched control experiments for evaluating the effectiveness of mathematics curricula, reading level might not be considered to be a relevant variable. However, in the case of the reform curricula that require a large amount of reading and a great deal of writing in stating the questions and producing results, differences in reading level may contribute significantly to the variance observed (Sconiers et al., 2002).

The goal of a comparative research design in establishing the effectiveness of a particular curriculum is to describe the net effect of a curricular program by estimating the gross outcome for an intervention group and subtracting the outcome for the comparable control group, while considering the design effects (contributed by the research methods) and stochastic effects (measurement fluctuations attributable to chance). To attribute cause and effect to curricular programs, one seeks a reasonable measure of the "counterfactual results," which are the outcomes that would have been obtained if the subjects had not participated in the intervention. Quasi-experimental methods seek a way to estimate this that involve probabilities and research design considerations. An evaluator also must work tenaciously to eliminate other likely factors that might have occurred simultaneously from outside uncontrolled sources.

We identified seven critical elements of comparative studies. These were: (1) design selection: experimental versus quasi-experimental, (2) methods of establishing comparability across groups, (3) selection of comparative units of analysis, (4) measures of implementation fidelity, (5) choices and treatment of outcomes, (6) selection of statistical tests, and (7) limits or constraints to generalizability. These design decisions are discussed in more detail, in relation to the actual studies, in Chapter 5, where the comparative studies that were analyzed for this report are reviewed.

Case Studies

Other evaluations focused on documenting how program theories and components played out in a particular case or set of cases. These studies, labeled case studies, often articulated in detail the complex configuration of factors that influence curricular implementation at the classroom or school level. These studies relied on the collection of artifacts at the relevant sites, interviews with participants, and classroom observations. Their goals included articulating the underlying mechanisms by which curricular materials work more or less effectively and identifying variables that may be overlooked by studies of less intensity (NRC, 2002). Factors that are typically investigated using such methods include understanding how faculties work together on decision making and implementation, or how attendance patterns affect instruction, or how teachers modify pedagogical techniques to fit their context, the preferences, or student needs.

The case study method (Yin, 1994, 1997) uses triangulation of evidence from multiple sources, including direct observations, interviews, documents, archival files, and actual artifacts. It aims to include the richness of the context; hence its proponents claim, "A major technical concomitant is that case studies will always have more variables of interest than data points, effectively disarming most traditional statistical methods, which demand the reverse situation" (Yin and Bickman, 2000). This approach also clearly delineates its expectations for design, site selection, data collection, data analysis, and reporting. It stresses that a slow, and sometimes agonizing, process of analyzing cases provides the detailed structure of argument often necessary to understand and evaluate complex phenomena. In addition to documenting implementation, this methodology can also include pre- and post outcome measures and the use of logic models, which, like program theory, produces an explicit statement of the presumed causal sequence of events in the cause and effect of the intervention. Because of the use of smaller numbers of cases, evaluators often can negotiate more flexible uses of open-ended multiple tests or select systematic variants of implementation variables.

Sharing of other features of the case study in relation to depth of placement in context and use of rich data sources is the *ethnographic evaluation*. Such studies may be helpful in documenting cases where a strong clash in values permeates an organization or project or where a cultural group may experience differential effects because their needs or talents are typical (Lincoln and Guba, 1986).

Other Considerations

Evaluator Independence

The relationship of an evaluator to a curriculum's program designers and implementers needs to be close enough to understand their goals and challenges, but sufficiently independent to ensure fairness and objectivity. During stages of formative assessment, close ties can facilitate rapid adjustments and modifications to the materials. However, as one reaches the stage of summative evaluation, there are clear concerns about bias when an evaluator is too closely affiliated with the design team.

Time Elements

The conduct of summative evaluations for examining curricular effectiveness must take into account the timeline for development, pilot testing, field testing, and subsequent implementation. Summative evaluation should be conducted only after materials are fully developed and provided to sites in at least field test versions. For curricula that are quite discontinuous with traditional practice, particular care must be taken to ensure that adequate commitment and capacity exists for successful implementation and change. It can easily take up to three years for a dramatic curricular change to be reliably implemented in schools.

Accumulation of Knowledge and the Meta-Analysis

For the purposes of this review of the evaluations of the effectiveness of specific mathematics curriculum materials, it is important to comment on studies that emphasize the accumulation of knowledge and meta-analysis. Lipsey (1997) persuasively argues that the accumulation of a knowledge base from evaluations is often overlooked. He noted that evaluations are often funded by particular groups to provide feedback on their individual programs, so the accumulation of information from program evaluation is left to others and often neglected. Lipsey (p. 8) argued that the accumulation of program theories across evaluations can produce a "broader intervention theory that characterizes . . . and synthesizes information gleaned from numerous evaluation studies." This report itself constitutes an effort to synthesize the information gleaned from a number of evaluation studies in order to strengthen the design of subsequent work. A further discussion of synthesis studies can be found in Chapter 6.

Meta-analysis produces a description of the average magnitude of effect sizes across different treatment variations and different samples. Based on the incomplete nature of the database available for this study, we decided that a full meta-analysis of program effects was not feasible. In addition, the extent of the variation in the types and quality of the outcome measures used in these studies of evaluating curricula makes effect sizes a poor method of comparison across studies. Nonetheless, by more informally considering effect size, statistical significance and the distribution of results across content strands, and the effects on various subgroups, one can identify consistent trends, evaluate the quality of the methodologies, and point to irregularities and targets for closer scrutiny through future research or evaluation studies.

These results also suggest significant implications for policy makers. Cordray and Fischer (1994, p. 1174) have referred to the domain of program effects as a "policy space" where one considers the variables that can be manipulated through program design and implementation. In Chapter 5, we discuss our findings in relation to such a policy space and presume to provide advice to policy makers on the territory of curricula design, implementation, and evaluation.

Our approach to the evaluation of the effectiveness of mathematics curricula seeks to recognize the complexity of the process of curricular design and implementation. In doing so, we see a need for multiple methodologies that can inform the process by the accumulation and synthesis of perspective. As a whole, we do not prioritize any particular method, although individual members expressed preferences. One strength of the committee was its interdisciplinary composition, and likewise, we see the determination of effectiveness as demanding the negotiation and debate among qualified experts.

For some members of the committee, an experimental study was preferred because theoretical basis for randomized or "controlled" experiments as developed by R. A. Fisher is a large segment of the foundation by which the scientific community establishes causal inference. Fisher invented the tool so that the result from a single experiment could be tested against the null hypothesis of chance differences. Rejecting the hypothesis of chance differences is probabilistically based and therefore runs the risk of committing a Type I error. Although a single, well-designed experiment is valuable, replicated results are important to sustain a causal inference, and many replications of the same experiment make the argument stronger. One must keep in mind that causal inference decisions are always probabilistically based.

Furthermore, it is important to remember that randomization is only a necessary but not a sufficient, condition for causal attribution. Other required conditions include the "controlled" aspect of the experiment, meaning that during the course of the experiment, there are no differences other than the treatment involved. The latter is difficult to ensure in natural school settings. But that does not negate the need for randomization because no matter how many theory-based variables can be controlled, one cannot refute the argument that other important variables may have been excluded.

The power of experimental approaches lies in the randomization of assignment to experimental and control conditions to avoid unintended but systematic bias in the groups. Proponents of this approach have demonstrated the dangers of a quasi-experimental approach in studies such as a recent one by Agodini and Dynarski (2001), which showed that a method of matching called "propensity studies" used in quasi-experimental design showed different results than an experimental study.

Others (Campbell and Stanley, 1966, pp. 2-3) developed these quasi-experimental approaches noting that many past experimentalists became disillusioned because "claims made for the rate and degree of progress which would result from experiment were grandiosely overoptimistic and were accompanied by an unjustified depreciation of nonexperimental wisdom." Furthermore, Cook (in press) acknowledged the difficulties associated with experimental methods as he wrote, "Interpreting [RCTs] results depends on many other things, an unbiased assignment process, adequate statistical power, a consent process that does not distort the populations to which results can be generalized, and the absence of treatment-correlated attrition, resentful demoralization, treatment seepage and other unintended products of comparing treatments. Dealing with these matters requires observation, analysis, and argumentation." Cook and Campbell and Stanley argue for the potential of quasi-experimental and other designs to add to the knowledge base.

In making these arguments, though experimentalists argue that experimentalism leads to unbiased results, their argument rests on idealized conditions of experimentation. As a result, they cannot actually estimate the level of departure of their results from such conditions. In a way, they thus leave the issues of external validity outside their methodological umbrella and largely up to the reader. Then, they tend to categorize other approaches in terms of how well they approximate their idealized stance. In contrast, quasi-experimentalists or advocates of other forms of method (case study, ethnography, modeling approaches) admit the questions of external validity to their designs, and rely on the persuasiveness of the relations among theory, method, claims, and results to warrant their conclusions. They forgo the ideal for the possible and lose a measure of internal validity in the process, preferring a porous relationship.

A critical component of this debate also has to do with the nature of cause and effect in social systems, especially those in which feedback is a critical factor. Learning environments are inevitably saturated with sources of feedback from the child's response to a question, to the pressures of high-stakes tests on curricular implementation. One can say with equal persuasion that use of a particular set of curricular materials caused the assignment of a student's score, which caused the student to learn the material in a curriculum. Cause and effect is best used to describe events in a temporal sequence where results can be proximally tied to causes based on the elimination of other sources of effect.

It is worth pointing out that the issues debated by members of this committee are not new, but have a long history in complex fields where the limitations of the scientific method have been recognized for a long time. Ecology, immunology, epidemiology, and neurobiology provide plenty of examples where the use of alternative approaches that include dynamical systems, game theory, large-scale simulations, and agent-based models have proved to be essential, even in the design of experiments. We do not live on a fixed landscape and, consequently, any intervention or perturbation of a system (e.g., the implementation of new curricula) can alter the landscape. The fact that researchers select a priori specific levels of aggregation (often dictated by convenience) and fail to test the validity of their results to such choices is not only common, but extremely limiting (validity).

In addition, we live in a world where knowledge generated at one level (often not the desired level) must be used to inform decisions at a higher level. How one uses scientifically collected knowledge at one level to understand the dynamics at higher levels is still a key methodological and philosophical issue in many scientific fields of inquiry. Genomics, a highly visible field at the moment, offers many lessons. The identification of key genes (and even the mapping of the human genome) is not enough to predict their expression (e.g., cancers) or to have enough knowledge that will help us to regulate them (e.g., cures to disease). "Nontraditional methods" are needed to make this fundamental jump. The evaluation of curricula successes is not a less complex enterprise and no single approach holds the key.

The committee does not need to solve these essential intellectual debates in order to fulfill its charge; rather, it chose to put forward a framework that could support an array of methods and forms of inference and evidence.

4

Content Analysis

Crucial to any curriculum is its content. For purposes of this evaluation, an analysis of the content should address whether the content meets the current and long-term needs of the students. What constitutes the long-term needs of the students is a value judgment based on what one sees as the proper goals and objectives of a curriculum. Differences exist among well-intentioned groups of individuals as to what these are and their relative priorities. Therefore, an analysis of a curricular program's content will be influenced by the values of the person or persons conducting the content analysis. Moreover, if the analysis considers a district, state, or national set of standards, other differences can be expected to emerge. In this chapter, we examine how to conduct the content analysis in order to identify a set of common dimensions that may help this methodology to mature, as well as to bring forth the difference in values. A curriculum's content must be compatible with all students' abilities, and it must consider the abilities of, and the support provided to, teachers.

An analysis of a curriculum's content should extend beyond a mere listing of content to include a comparison with a set of standards, other textual materials, or other countries' approaches or standards. For the purposes of this study—reviewing the evaluations of the effectiveness of mathematics curricula—content analyses will refer to studies that range from documenting the coverage of a curriculum in relation to standards to more extended examinations that also assess the quality of the content and presentation. Clarity, consistency, and fidelity to standards and their relationship to assessment should be clearly identifiable, basic elements of any

TABLE 4-1 Distribution of the Content Analysis Studies: Studies by Type and Reviews by Grade Band

Type of Study	Number of Reviews	Number of Studies	Percentage of Total Studies by Program Type
NSF supported		19	53
Elementary	10		
Middle	20		
High	13		
Total	43		
Commercially generated		1	3
Elementary	3		
Middle	8		
High	0		
Total	11		
UCSMP		10	28
UCSMP (high school)	12		
Total	12		
Not counted in above		6	16
Total		36	100

reasonable content analysis. The remainder of the chapter reviews primary examples of content analysis and delineates a set of dimensions for review that might help make the use of content analysis evaluations more informative to curriculum decision makers.

We identified and reviewed 36 studies of content analysis of the supported and commercially generated National Science Foundation (NSF) mathematics curricula. Each study could include reviews of more than one curriculum. Table 4-1 lists how many studies were identified in each program type (NSF-supported, University of Chicago School Mathematics Project [UCSMP], and commercially generated), the total number of reviews in those studies, and the breakdown of those reviews by elementary, middle, or high school. These reviews allowed us to consider various approaches to content analysis, to explore how those variations produced different types of results and conclusions, and to use this information to make inferences about the conduct of future evaluations.

The content analysis reviews were spread across the 19 curricular programs under review. The number of reviews for each curricular program varied considerably (Table 4-2); hence our report on these evaluations draws on reports by some programs more than others.

Table 4-3 identifies the sources of the studies by groups that produced multiple reviews. Those classified as internal were undertaken directly by an author, project evaluator, or member of the staff of a publisher associated with the curricular program. Content analyses categorized as external

TABLE 4-2 Content Counts by Program

Curriculum Name	0	1	2-5	>5
			Number of Reviews by Program	
Everyday Mathematics				7
Investigations in Data, Number and Space			2	
Math Trailblazers		1		
Connected Mathematics Project				8
Mathematics in Context				7
Math Thematics			5	
MathScape		1		
MS Mathematics Through Applications Project	0			
Interactive Mathematics Project			4	
Mathematics: Modeling Our World (ARISE)			2	
Contemporary Mathematics in Context (Core-Plus)			3	
Math Connections			2	
SIMMS			2	
Addison Wesley/Scott Foresman			2	
Harcourt Brace		1		
Glencoe/McGraw-Hill			2	
Saxon				6
Houghton Mifflin/McDougal Littell	0			
Prentice Hall/UCSMP				14
Total number of times a curriculum is in any study	68			
Number of evaluation studies	36			

were written by authors who were neither directly associated with the curricula nor with any of the sources of multiple studies.

In this chapter (and in subsequent chapters) we report examples of evaluation studies and describe their approaches and statements of findings. It must be recognized that the committee does not endorse or validate the accuracy and appropriateness of these findings, but rather uses a variety of them to illustrate the ramifications of different methodological decisions. We have carefully selected a diverse set of positions to present a balanced and fair portrayal of different perspectives. Knowing the current knowledge base is essential for evaluators to make progress in the conduct of future studies.

TABLE 4-3 Number of Reviews of Content by Program Type

	NSF	Commercial	UCSMP
AAAS	9	2	1
Author, external	17		1
Author, internal	5		11
Mathematically Correct	6	9	1
U.S. Department of Education	6		
TOTAL	43	11	14

NOTE: 11 of the 17 NSF authors, external are from three projects funded by NSF.

LITERATURE REVIEW

We identified four sources where systematic content analyses were conducted and applied across programs. They were (1) the U.S. Department of Education's review for promising and exemplary programs; (2) the American Association for the Advancement of Science (AAAS) curricular material reviews for middle grades mathematics and algebra under Project 2061; (3) the reviews written by Robinson and Robinson (1996) on the high school curricula supported by the National Science Foundation; and (4) the reviews of grades 2, 5, 7, and algebra, available on the Mathematically Correct website. We begin by reviewing these major efforts and identifying their methodology and criteria.

U.S. Department of Education

The U.S. Department of Education's criteria for content analysis evaluation—Quality of Program, Usefulness to Others, Educational Significance, and Evidence of Effectiveness and Success (U.S. Department of Education, 1999)—included ratings on eight criteria structured in the form of questions:

1. Are the program's learning goals challenging, clear, and appropriate; is its content aligned with its learning goals?
2. Is it accurate and appropriate for the intended audience?
3. Is the instructional design engaging and motivating for the intended student population?
4. Is the system of assessment appropriate and designed to guide teachers' instructional decision making?
5. Can it be successfully implemented, adopted, or adapted in multiple educational settings?
6. Do its learning goals reflect the vision promoted in national standards in mathematics?

7. Does it address important individual and societal needs?

8. Does the program make a measurable difference in student learning?

American Association for the Advancement of Science

AAAS's Project 2061 (http://www.project2061.org) developed and used a methodology to review middle grades curricula and subsequently algebra materials. In describing their methods, Kulm and colleagues (1999) outlined the training and method. After receiving at least three days of training before conducting the review, each team rated a set of algebra or middle grades standards in reference to the Curriculum and Evaluation Standards for School Mathematics from the National Council of Teachers of Mathematics (NCTM) (1989). In algebra, each review encompassed ideas from one area of algebra: functions, operations, and variables. Similarly in middle grades, particular topics were specified. Each team used the same idea set to review a total of 12 textbooks or sets of textbooks over 12 days. Two teams reviewed each book or set of materials.

The content analysis procedure is outlined on the AAAS website and summarized below:

• Identify specific learning goals to serve as the intellectual basis for the analysis, particularly to select national, state, or local frameworks.

• Make a preliminary inspection of the curriculum materials to see whether they are likely to address the targeted learning goals.

• Analyze the curriculum materials for alignment between content and the selected learning goals.

• Analyze the curriculum materials for alignment between instruction and the selected learning goals. This involves estimating the degree to which the materials (including their accompanying teacher's guides) reflect what is known generally about student learning and effective teaching and, more important, the degree to which they support student learning of the specific knowledge and skills for which a content match has been found.

• Summarize the relationship between the curriculum materials being evaluated and the selected learning goals.

The focus in examining the learning goals is to look for evidence that the materials meet the following objectives:

• Have a sense of purpose.
• Build on student ideas about mathematics.
• Engage students in mathematics.
• Develop mathematical ideas.
• Promote student thinking about mathematics.

- Assess student progress in mathematics.
- Enhance the mathematics learning environment.

Features of AAAS's content analysis process include opting for a careful review of a few selected topics, chosen prior to the review, that are used consistently across the textbooks reviewed, and requiring that particular page numbers and sections are referenced throughout the review. The evaluation reviews are available on the website.

Robinson and Robinson

Robinson and Robinson (1996) reviewed the integrated high school curricula by defining and using a set of threads to review a set of integrated curricula. These threads, which included algebra/number/function, geometry, trigonometry, probability and statistics, logic/reasoning, and discrete mathematics, were effective in bringing together the structure of an integrated curricula. They further identify commonalities among the curricula, including what is meant by integrated and context-rich, what pedagogies are used, choices of technology, and methods of assessment.

Mathematically Correct

On the Mathematically Correct website, reviews of curricular materials are made available, often via links to other sites viewed as similarly aligned; we collected those pertaining to the relevant curricula. A set of reviews of 10 curricular programs for grades 2, 5, and 7 by Clopton and colleagues presented a systematic discussion of the methodology where the reviewers selected topics "designed to be sufficient to give a clear impression of the features of the presentation and an assessment of the mathematical depth and breadth supported by the program" (1999a, 1999b, 1999c). Ratings on programs focused on mathematical depth, the clarity of objectives, and the clarity of explanations, concepts, procedures, and definitions of terms. Other foci included the quality and sufficiency of examples and the efficiency of learning. In terms of student work, the focus was on the quality and sufficiency of student work and its range, depth, and scope. Clopton et al.'s studies used the Mathematically Correct and the San Diego Standards that were in force at the time. For example, for fifth grade, they selected multiplication and division of whole numbers, decimal multiplication and division, area of triangles, negative numbers and powers, exponents, and scientific notation where a detailed rationale is provided for the selection of each topic. Each textbook was rated according to the depth of study of the topics, ranging from 1 (poor) to 5 (outstanding). Two dimensions of that judgment were identified: quality of the presentation (clarity of objectives, explanations, examples, and efficiency of learning) and quality of student work (sufficiency, range, and depth). No procedure was outlined for how reviewers were trained or how differences were resolved.

Other Sources of Content Analysis

The rest of the studies categorized as content analyses varied in the extent of specificity on methodology. A number of the reviews were targeted directly at teachers to assist them in curricular decision making. The Kentucky Middle Grades Mathematics Teacher Network (Bush, 1996), for example, reviewed four middle grades curricula using groups of teachers. They selected four general content areas (number/computation, geometry/measurement, probability/statistics, and algebraic ideas) from the Core Content for Assessment (Kentucky Department of Education, 1996) and asked teachers to evaluate the materials based on appropriateness for grade levels and quality of content presentation. They were also asked to evaluate the pedagogy. Although these reviews identified missing content strands and produced judgments of general levels of quality, we found them to be of limited rigor for use in our study; in particular, this was because of their lack of specificity on method. Other content analyses ranged from those by authors explaining the design and structure of the materials (Romberg et al., 1995) to those by critics of particular programs focusing on only the areas of concerns, often with sharp criticisms.

We identified one study by Adams et al. (2000) (subsequently referred to as the "Adams report") entitled *Middle School Mathematics Comparisons for Singapore Mathematics, Connected Mathematics Program, and Mathematics in Context* that used the National Council of Teachers of Mathematics (NCTM) Principles and Standards for School Mathematics 2000 as a comparison. Their method of content analysis is based on 72 questions that compare the curricula against the 10 overarching standards (number, algebra, geometry, measurement, data and probability, problem solving, reasoning and proof, communication, connection, and representation) and 13 questions that examine six principles (equity, curriculum, teaching, learning, assessment, and technology). Further information from each of these four sources of content analysis will be referenced in the discussion of the reviews in subsequent sections.

DIMENSIONS OF CONTENT ANALYSES

As we discovered by examining the available reports, the art and science of systematic content analysis for evaluation purposes is still in its adolescence. There is a clear need for the development of a more rigorous paradigm for the planning, execution, and evaluation of content analyses.

This conclusion is supported by a review of the results of the content analyses; the ratings of many curricular programs vacillated from strong to weak, with little explanation for the divergent views.

In Chapter 3, we identified content analyses as a form of connoisseurial assessment (Eisner, 2001). Variations among experts can be expected, but they should be connected to differences in selected standards, philosophies of mathematics or learning, values, or selection of topics for review. Each of these influences on the content analysis should be explicitly articulated to the degree possible as part of an evaluation report, a form of "full disclosure" of values and possible bias. Faced with the current status of "content analysis," we recognized that rather than specifying methods for the conduct of content analyses to produce identical conclusions, we needed to assist evaluators in providing decision makers with clearer delineation of the different dimensions and underlying theories they used to make their evaluations.

A review of the research literature provided relatively few systematic studies of content analysis. However, we did find three studies that were applicable to our work: (1) an international comparative analysis and distribution of mathematics content curricula across the school grades that was conducted in the Third International Mathematics and Science Study (TIMSS) (Schmidt et al., 2001), (2) a project by Porter et al. (1988) concerning content determinants where they demonstrated variations in emphases in a variety of textbooks in the early 1980s, and (3) where Blank (2004) developed tools to map the content dimensions of curriculum and compared this with the relative emphases in the assessments.

To strengthen our deliberations on this issue, we collected testimony from educators and invited scholars to comment on what constitutes a useful content analysis, and illustrations are cited in text boxes. Our review of the literature, analysis of the submitted evaluations, and consideration of the responses confirmed our belief that uniform standards or even a clear consensus on what constitutes a content analysis do not exist. We saw our work as a means to contribute to the development of clearer methodological guidelines through a synthesis of previous studies and the deliberations of the committee. In the next sections, we discuss participation in content analyses, the selection of standards or comparative curricula, and the inclusion of content and/or pedagogy. We then identify a set of dimensions of content analyses to guide their actual conduct.

Participation in Content Analyses

A key dimension of content analysis is the identity of the reviewer(s) (Box 4-1). We recognized the importance of including members of the mathematical sciences community in this process, including those in both

---

BOX 4-1
Comments on Participation in Content Analysis

"Whether or not a curriculum is mathematically viable would ultimately depend on the judgment of some mathematicians. As such, good judgment rightfully should play a critical role."—Hung Hsi Wu, University of California, Berkeley

"Neither mathematicians nor educators are alone qualified to make these judgments; both points of view are needed. Nor can the work be divided into separate pieces (the intended and achieved curriculum), one for mathematicians to judge and one for educators. The two groups must cooperate."—William McCallum, University of Arizona

"The qualifications needed to make valid judgments on the quality of content in terms of focus, scope, and sequencing over the years are much more stringent than seems to be commonly appreciated. In mathematics, for example, someone with *only a Ph.D. in the subject is unlikely to be qualified* to do this unless, over the years, he or she has made significant contributions in many areas of the subject and has worked successfully in applying mathematics to outside subjects as well."—R. James Milgram, Stanford University

---

pure and applied mathematics (Wu and Milgram testimony at the September 2002 Workshop). There was agreement among respondents that the expertise of mathematics educators is needed to ensure careful considerations of student learning and classroom practices (McCallum testimony at the September 2002 Workshop). Although we found some debates between the parties, our reviews also revealed that mathematicians and mathematics educators shared a common goal of an improved curriculum. Controversy in this direction provides increased impetus for establishing methodological guidelines and bringing to the surface the underlying reasons for disputes and disagreements.

We found it particularly helpful when a content analysis was accompanied by a clear statement of the reviewer's expertise. Adams et al. (2000, p. 2), for example, disclosed their qualifications early on, writing, "Another point to consider is our expertise, both what it is and what it is not. The group that created this comparison consists entirely of people who combine high-level training in mathematics with an interest in education but who have neither direct experience of teaching in the American K-12 classroom nor the training for it." Most reviews, however, did not discuss reviewers' credentials. We did not find, for example, a similar statement by experts in education identifying their qualifications in mathematics. Other reviews, such as the one by Bush (1996), were conducted by panels of teachers. The involvement of teachers and/or parents in content analysis should prove to be a valuable means of feedback to writers and a source of professional development to the teachers.

The Selection of Standards or Comparative Curricula

At the most general level, the conduct of a content analysis requires identifying either a set of standards against which a curriculum is compared or an explicitly contrasting curriculum; the analysis should not rely on an imprecise characterization of what should be included. Common choices include the original or the revised NCTM Standards, state standards or other standards, or comparative curricula as a means of contrast. The strongest evaluations, in our opinion, used a combination of these two approaches along with a rationale for their decisions.

The choice of comparison can have a crucial impact on the review. For example, the unpublished Adams report succinctly showed how conclusions from content analysis of a curriculum can vary with changes in the adopted measures, varying goals, and philosophies. This report, prepared for the NSF, stood out as being particularly complete and carefully researched and analyzed in its evaluations. To appraise the NSF curricula, they evaluated Connected Mathematic Project (CMP) and Mathematics in Context in terms of the 2000 NCTM Principles and Standards for School Mathematics. These two programs were chosen, in part, because evidence by AAAS's Project 2061 suggested they are among the top of the 13 NSF-sponsored projects studied.

An interesting and valued feature in the Adams report was that these programs were compared with the "traditional" Singapore mathematics textbooks "under the authority of a traditional teacher" (Adams et al., 2000, p. 1). To explain why Adams selected the Singapore program as the "traditional approach" measure of comparison, recall that the performance of students from the United States on TIMSS "dropped from mediocre at the elementary level through lackluster at the middle school level and down to truly distressing at the high school level." On the other hand, of the 41 nations whose students were tested, those from Singapore "scored at the very top" (Adams et al., 2000, p. 1). We found that this comparison component of the Adams study with a top-ranked traditional program provides a valuable and new dimension absent from most other studies. Because the United States is at the forefront of scientific and technological advances, the Singapore comparison dimension cannot be ignored: content analysis studies that make comparisons across a variety of types of curricular material must be encouraged and supported.

The Adams report demonstrated the importance of the selection of the standards and the comparative curricula in their reported results. When these NSF programs were compared to the NCTM Principles and Standards for School Mathematics (PSSM), they showed strong alignment; when contrasted with the Singapore curriculum, they revealed delays in the introduction of basic material.

Once the participants, standards, and comparison curricula are selected and identified, other factors remains that are useful in creating the necessary types of distinctions to conduct the content analyses. In the next section, we identify some of these additional considerations.

Inclusion of Content and/or Pedagogy

A major distinction among content analyses was whether the emphasis was on the material to be taught, or the material and its pedagogical intent. In response to criticisms of American programs, which are described as "a mile wide and an inch deep" (Schmidt et al., 1996), a focus on the identification and treatment of "essential ideas" provides one way to approach the content dimension of content analysis (Schifter testimony at the September 2002 Workshop). Another respondent proposed a three-part analysis to determine whether (1) there are clearly stated objectives to be achieved when students learn the material, (2) the objectives are reasonable, and (3) the objectives will prepare students for the next stages of learning (Milgram testimony at the September 2002 Workshop).

Alternatively, some scholars believe a content analysis should include an examination of pedagogy citing the need to examine both the content and how it is intended to be taught (Gutstein testimony at the September 2002 Workshop). Some scholars expressed the view that the content analysis must be conducted in situ in recognition that the content of a curriculum emerges from the study of classroom interactions, so examination of materials alone is insufficient (McCallum testimony at the September 2002 Workshop). This approach is not simply a question of whether material was taught and how well; it reflects a concern that instructional practices can modify the course of the content development, especially when using activity-based approaches that rely on the give and take of classroom interactions to complete the curricula. Such studies, often called design experiments (Cobb et al., 2003), do not entail full-scale implementation, only pilot sights or field tests. Typically, these student-oriented approaches rely more heavily on teachers' content knowledge and judgment and hence are difficult to analyze in the absence of their use and without information about the teacher's ability. Decisions to include or exclude pedagogy in a content analysis and to study the materials independently or *in situ* are up to the reviewer, but the choice and reasons for it should be specified. For examples of viewpoints from experts, see Box 4-2.

BOX 4-2
Major Considerations in Content Analysis

"A useful content analysis would . . . identify the essential mathematical ideas that underlie those topics . . . [one must determine if] those concepts [are] developed through a constellation of activities involving different contexts and drawing upon a variety of representations, and so one must follow the development of those concepts through the curriculum. . . . A curriculum should include enough support for teachers to enact it as intended. Such support should allow teachers to educate themselves about mathematics content, students' mathematical thinking, and relevant classroom issues. . . . It might help . . . teachers analyze common student errors in order to think about next steps for those who make them. And it might help teachers figure out how to adjust an activity to make it more accessible to struggling students without eliminating the significant mathematics content, or to extend the activity for those ready to take on extra challenge."—Deborah Schifter, Education Development Center

"To begin, one must understand that mathematics is almost unique among the subjects in the school curriculum in that what is done in each grade depends crucially on students having mastered key material in previous grades. Without this background, students simply cannot develop the understanding of the current year's material to a sufficient depth to support future learning. Once this failure happens, students typically start falling behind and do not recover."—R. James Milgram, Stanford University

"Mathematics is by definition a coherent logical system and therefore tolerates no errors. Any mathematics curriculum should in principle be error-free."—Hung Hsi Wu, University of California, Berkeley

[A content analysis requires] "looking both at the mathematics that goes into it and at how well the mathematics takes root in the minds of students. Both are necessary: topics that are mentioned in the table of contents but unaccompanied by a realistic plan for preparing teachers and reaching students cannot be said to add to the content of a curriculum, nor can activities that work well in the classroom but are mathematically ill conceived."—William McCallum, University of Arizona

"One very important [issue of vital importance to mathematics education that is not captured by broad measures of student achievement] is the maintenance of a challenging high-quality education for the best students. The primary object of concern these days in mathematics education seems to be the low-achieving student, how to raise the floor. Certainly, this is the spirit evoked by No Child Left Behind. This is a very important issue, but it should not blind us to the fact that the old system, the system often denigrated today, is the one which got us where we are, to a society transformed by the impact of technology. . . . The percentage of people who need to be highly competent in mathematics has always been and will continue to be small, but it will not get smaller. We must make sure that mathematics education serves these people well."—Roger Howe, Yale University

The Discipline, the Learner, and the Teacher as Dimensions of Content Analysis

Curricular content analysis involves an examination of the adequacy of a set of materials in relation to its treatment of the discipline, learner(s), and the teacher. Learning occurs through an interaction of these three elements. Beyond being examined independently, an analysis must consider how they interact with one another. Following this triadic scheme, we specified three dimensions of a content analysis to be more informative to curricula decision makers:

1. Clarity, comprehensiveness, accuracy, depth of mathematical inquiry and mathematical reasoning, organization, and balance (disciplinary perspectives).
2. Engagement, timeliness and support for diversity, and assessment (learner-oriented perspectives).
3. Pedagogy, resources, and professional development (teacher- and resource-oriented perspectives).

We discuss each dimension and how it has been addressed in the various content analyses under review. Elements of the dimensions overlap and interact with each other, and no dimension should be assumed as logically or hierarchically prior to the others, except as indicated within a particular content analysis. A content analysis neglecting any dimension would be considered incomplete.

Dimension One: Clarity, Comprehensiveness, Accuracy, Depth of Mathematical Inquiry and Mathematical Reasoning, Organization, and Balance

A major goal of developing standards is to provide guidance in the evaluation of curricular programs. Thus, in determining curricular effectiveness at a basic level, one wants to ascertain if all the relevant topics are covered, if they are sequenced in a logical and coherent structure, if there is an appropriate balance and emphasis in the treatment, and if the material appropriately meets the longer term needs of the students. The many ways to do this means there is a significant element of judgment in making this assessment. Nonetheless, it is likely that some curricula do a better job than others, so distinctions must be drawn.

In developing a method to determine these factors, many content analyses found that determining comprehensiveness can be difficult if curricular programs offer too many disjointed and overlapping topics. This finding motivated a call for the **clarity** of objectives or the identification of the major conceptual ideas (Milgram testimony; Schifter testimony). With a clear specification of objectives, a reviewer can search for missing or superfluous content. A common weakness among content analyses, for example, was failure to check for **comprehensiveness**. In mathematics, comprehensiveness is particularly important, as missing material can lead to an inability to function downstream. In a content analysis of the Connected Math Program, Milgram (2003) wrote:

> Overall, the program seems to be very incomplete, and I would judge that it is aimed at underachieving students rather than normal or higher achieving students. . . . The philosophy used throughout the program is that students construct their own knowledge and that calculators are to always be available for calculation.

This means that

- standard algorithms are not introduced, not even for adding, subtracting, multiplying, and dividing fractions
- precise definitions are never given
- repetitive practice for developing skills, such as basic manipulative skills, is never given

Likewise, Adams et al. (2000, p. 14), critiqued Mathematics in Context on similar dimensions:

> Our central criticism of Mathematics in Context curriculum concerns its failure to meet elements of the 2000 NCTM number strands. Because MiC is so fixated on conceptual underpinnings, computational methods and efficiency are slighted. Formal algorithms for, say, dividing fractions are neither taught nor discovered by the students. The students are presented with the simplest numerical problems, and the harder calculations are performed using calculators. Students would come out of the curriculum very calculator-dependent. . . . To us, this represents a radical change for the old "drill-and-kill" curricula, in which calculation was over-emphasized. The pendulum has, apparently, swung to the other side, and we feel a return to some middle ground emphasizing both conceptual knowledge and computational efficiency is warranted.

As an example of the positive impact of content analyses, the authors have indicated that in response to criticisms and the changes advised by PSSM, plans are being made to strengthen these dimensions in subsequent versions (Adams et al., 2000).

Accuracy was selected as one of our primary criteria because all consumers of mathematics curricula expect and demand it. The elimination of errors is of critical importance in mathematics (Wu testimony at the September 2002 Workshop). Some content analyses claimed that materials had too many errors (Braams, 2003a). Virtually no one disputes that curricular materials should be free from errors; all authors and publishers indicated that errors should be quickly corrected, especially in subsequent versions. It appears that not all content analyses paid appropriate attention to the accuracy issue. For example, Richard Askey, University of Wisconsin, in commenting on the Department of Education reviews to the committee during his September 17, 2002, testimony, pointed out that "In these 48 reviews, no mention of any mathematical errors was made. While a program could be promising and maybe even exemplary and contain a significant number of mathematical errors, the fact that no errors were mentioned strongly suggests that these reviews were superficial at best."

It surfaced over time that some of the debate over the quality of the materials focused on the relative importance of different types of mathematical activity. To assist in deliberations, we chose to stipulate a distinction between **mathematical inquiry** and **mathematical reasoning**. Mathematical inquiry, as used in the report, refers to the elements of intuition necessary to create insight into the genesis and evolution of mathematical ideas, to make conjectures, to identify and develop mathematical patterns, and to conduct and study simulations. Mathematical reasoning refers to formalization, definition, and proof, often based on deductive reasoning, formal use of induction, and other methods of establishing the correctness, rigor, and precise meaning of ideas and patterns found through mathematical inquiry. Both are viewed as essential elements of mathematical thought, and often interact. Making too strong a distinction between these two elements is artificial.

Frequent debates revolve around the balance between mathematical inquiry and mathematical reasoning. For example, when content material has weak or poor explanations, does not establish or is not based on appropriate prerequisites, or fails to be developed to a high level of rigor or lacks practice in effective choices of examples, issues of mathematical reasoning are often cited as missing. At the same time, rather than focusing solely on the treatment of a particular topic at one particular point in the material, it is essential to follow the entire trajectory of conceptual development of an idea, beginning with inquiry activities, and ensuring that the subsequent necessary formalization and mathematical reasoning are provided. Moreover, one must determine in a content analysis whether a balance between the two is achieved so that the material both invites students' entry and exploration of the origin and evolution of the ideas and builds intuition, and ensures their development of disciplined forms of evidence and proof.

To illustrate this tension and the need for careful communication and exchange around these issues, we report here two viewpoints, one pre-

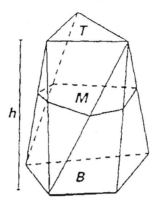

FIGURE 4-1 Prismoidal figure.
SOURCE: Contemporary Mathematics in Context: A Unified Approach (Core-Plus), Course 2, Part A, p. 137.

sented orally to the committee at one of the workshops and the other subsequently sent to the committee. Richard Askey (University of Wisconsin), who has examined curricular adequacy, responded at the Workshop to our question as to whether the program's content reflects the nature of the field and the thinking that mathematicians use. He discussed the importance of careful examination of content for errors and inadequate attention to rigor and adequacy, focusing on Core-Plus curriculum's treatment of formula for the volume of a number of common three-dimensional shapes. In Core-Plus, 9th-grade students are introduced to a table of values that represent the volume of water contained in a set of shapes (including a square pyramid, a triangular pyramid, a cylinder, and a cone) as the height increases from 0 to 20 cm. Using the data and a calculator, the students fit a line to produce the inductive observation that the volume of a cone is approximately one-third the volume of a cylinder with identical height and base. Askey's stated objection was "The geometric reasoning for the factor of 1/3 [is] never mentioned in Core-Plus."

A follow-up discussion of volume is presented on the next page, where the authors provide a formula for a prismoidal 3-D figure (Figure 4-1). To calculate its volume, a formula is given:

$$V = \left(\frac{B + 4M + T}{6} \right) h$$

In the formula, B is the area of the cross-section at the base, M is the area of the cross-section at the "middle," T is the area of the cross-section at the top, and h is the height of the space shape.

The authors of Core-Plus informally explain this formula as a form of a weighted average, then use it to derive the formula for the particular cases of a triangular prism, a cylinder, a sphere, and a cone. They end with problems asking students to estimate volumes of vases or bottles.

Askey's objections were stated as:

> A few pages later, a result which was standard in high school geometry books of 100 years ago and of some as late as 1957 (with correct proofs) is partially stated, but no definition is given for the figures the result applies to, nothing is written about why the result is true, and it is applied to a sphere. The correct volume of sphere is found by applying this formula, but the reason for this is not that the formula as stated applies to a sphere, but because the formula is exact for any figure where the cross-sectional area is a quadratic function of the height, which is the case for a sphere as well as the figures which should have been defined in the statement of the prismoidal formula. To really see that the use of this formula for a sphere is wrong, consider the two-dimensional case. The corresponding formula is a formula for the area of a trapezoid, and one does not get the right formula for the area of a circle by applying a formula for the area of a trapezoid, unless one uses an integral formula for area. The prismoidal formula is just Simpson's rule applied to an integral. The idea that you can apply formulas at will is not how mathematicians think about mathematics.

The committee received a letter in response to Askey's assertions from Eric Robinson (Department of Mathematics, Ithaca College), who wrote:

> One panelist used two examples for the Core-Plus curriculum in his presentation to suggest a general lack of reasoning in the curricula. These examples had other intents. The criticism in the first example has to do with the lack of geometric derivations of the factor of 1/3 in the formula for the volume of a cone. . . . In Core-Plus, when students explore examples, they collect "data." They have studied some of the statistics strand earlier in the curriculum, where, among other things, they look for patterns in data. Here, and in other places in the curriculum, there is a nice tie between the idea of statistical prediction and conjecture. Students would understand the difference between a conjecture and a result that was certain. They wouldn't interpret it as being "able to do anything they want". The intent here is to tie together some ideas the students have been studying; patterns in data, linear models and volume. It is also intended to suggest a geometric interpretation of the coefficient of x in the symbolic representation of the linear model. It is not intended to be derivation of formulas for any of the solids mentioned. . . . The second example mentions references on the prismoidal formula. The intent of that particular

exercise is to see if students can translate information from a visual setting to a symbolic expression. The intent, here, is not the derivation of the cone or prismoidal formulas. Intent matters.

These differing views about content provide a valuable window into the challenges associated with the evaluation of content analyses. In the committee's view, Askey seeks to have more attention paid to issues of mathematical reasoning and abstraction: formal introduction, complete specification of restriction of formula application, and proof. Robinson values mathematical inquiry: conjecturing based on data, the development of visual reasoning, statistical prediction, and successive thematic approaches to strengthen intuition. Robinson further argues against isolated review of the treatment of topics because "students are continually asked to provide justification, explanation, and evidence, etc., in various ways that focus on obtaining insight needed to answer a question or solve a problem. Student justification increases in rigor as students ascend through the grades."

The example illustrates that content analysts can reasonably have different views about the optimal balance of mathematical reasoning and inquiry, and the debates over content have mathematically and pedagogically sophisticated arguments on both sides. We encourage discussions in line with the differing Askey and Robinson views, at the level of particular example and with careful and explicit articulation of reason and logic. More precise methodological distinctions can be useful in facilitating such exchanges.

Examining the **organization** of material is another part of content analysis. For some, an analysis of the logical progression of concept development determines the sequence of activities, usually in the form of a Gagne-type hierarchical structure (Gagne, 1985). In reviewing Saxon's Math 65 treatment of multiplication of fractions, Klein (2000) illustrates what is called an "incremental approach":

> Multiplication of fractions is explained in the Math 65 text using fraction manipulatives. Pictures in the text, beginning on page 324, help to motivate the procedure for multiplying fractions. This approach is perhaps not as effective as the area model used by Sadlier's text, however, it leads in a natural way to an intuitive understanding of division by fractions, at least for simple examples. On page 402 of the Math 65 text, the notion of reciprocals is introduced in a clear, coherent fashion. In the following section, fraction division is explained using the ideas of reciprocals as follows. On page 404, the fraction division example of 1/2 divided by 2/3 is given. This is at first interpreted to mean, "How many 2/3's are in 1/2?" The text immediately acknowledges that the answer is less than 1, so the question needs to be changed to "How much of 2/3 is in 1/2?" This question is supported by a helpful picture. The text then reminds the

students of reciprocals and reduces the problem to a simpler one. The question posed is, "How many 2/3's are in 1?" This question is answered and identified as the reciprocal, 3/2. The text returns to the original question of 1/2 divided by 2/3. It then argues that the answer should be 1/2 times the answer to 1 divided by 2/3. The development proceeds along similar well-supported arguments from this point. Although this treatment of fraction division is clearer than what one finds in most texts, it would be improved with further elaboration on the problem 1 divided by 2/3. Here, the inverse relationship between multiplication and division could be invoked with great advantage. 1 divided by 2/3 = 3/2 because $3/2 \times 2/3 = 1$.

Some curricula use a spiral or thematic approach, which involves the weaving of content strands throughout the material with increased levels of sophistication and rigor. In the Department of Education review (1999), Everyday Math was commended for its depth of understanding when a reviewer wrote, "Mathematics concepts are visited several times before they are formally taught. . . . This procedure gives students a better understanding of concepts being learned and takes into consideration that students possess different learning styles and abilities" (criterion 2, indicator b). In contrast, Braams (2003b) reviewed the same materials and wrote, "The Everyday Mathematics philosophical statement quoted earlier describes the rapid spiraling as a way to avoid student anxiety, in effect because it does not matter if students don't understand things the first time around. It strikes me as a very strange philosophy, and seeing it in practice does not make it any more attractive or convincing." We know of no empirical studies that shed light on the differences in these assertions of preference.

Finally, the issue of **balance** referred to the relative emphasis among choices of approaches used to attain comprehensiveness, accuracy, depth of mathematical inquiry and reasoning, and organization. Curricula are enacted in real time, and international comparisons show that American textbooks are notoriously long, providing the appearance of the ability to cover more content than is feasible (Schmidt et al., 1996). In real time, curricular choices must be made and curricular materials reflect those choices, either as written or as enacted. In reference to mathematics curricula, decisions on balance include: conceptual versus procedural, activities versus practice, applications versus exercises, and balance among selected representations such as the use of numerical data and tables, graphs, and equations.

The level of use and reliance on technology is another consideration of balance, and many content analyses comment directly on this. Analyses of curricular use of technologies fell into two categories: the use of calculators in relation to computations and the use of technologies for modeling and simulation activities, typically surrounding data use.

For some reviewers, the use of calculators is questioned for its effects on computational fluency. In summarizing their reviews of the algebra materials, Clopton et al. (1998) wrote:

> In general, the material in the bulk of the lessons would receive moderately high marks for presentation and student work, but many of the activities and the heavy use of calculators lead to serious concerns and a lowering of both presentation and student work subscores. In general, but not always, nearly every one of the books in this category would be improved substantially with removal of most discovery lessons and the over use of calculators.

In the review of UCSMP, the statement was made, "The use of technology does not generally interfere with learning, and there is a decided emphasis on the analytic approach" (Clopton et al., 1998).

Other reviewers value the use of technologies as a positive asset in a curriculum and make finer distinctions about appropriate uses. It is valuable to consider whether certain kinds of problems and topics would be eliminated from a curriculum without the use and access to appropriate technology, such as problems using large or complex databases. For example, although AAAS reviews (1999a) do not have a criteria addressing technology in isolation, they include it in a category referred to as "Providing Firsthand Experiences." In reviewing the Systemic Initiative for Montana Mathematics and Science (SIMMS) Integrated Mathematics: A Modeling Approach Using Technology materials, they write, "Activities include graphing tabular data, creating spreadsheets and graphs, collecting data using calculators, and comparing and predicting values of such things as congressional seat apportionment and nutritional values. Although the efficiency of doing some activities depends heavily on the teachers' comfort level with the technology, these firsthand experiences are efficient when compared to others that could be used." A discussion of the balance of the use of technology is a useful way for a reviewer to indicate his or her stance on the appropriate place of technology in the curriculum. In content analyses, we indicated that rather than making statements that indicated either complete rejection or unbridled acceptance of technological use, a more useful approach is to more clearly identify effects on particular mathematical ideas in terms of gains and losses. A fuller discussion of a range of technology beyond calculators should be included in content analyses.

Another issue of organization is determining how much direct and explicit instructional guidance to include in the text. Discovery-oriented or student-centered materials may choose to include less explicit content development in text, and rely more heavily on the tasks to convey the concepts. Although this provides the opportunity for student activity, it relies more heavily on the teachers' knowledge. In this sense, curriculum materials may vary between serving as a reference manual and a work or activity book. Perhaps it would resolve many of these disputes if both types of curricular resources were distinctively provided to schools as different types of resources.

As an example of debates on issues of balance, Adams et al. (2000, p. 12) wrote:

> CMP admits that "because the curriculum does not emphasize arithmetic computations done by hand, some CMP students may not do as well on parts of the standardized tests assessing computational skills as students in classes that spend most of their time on practicing such skills. This statement implies we have still not achieved a balance between teaching fundamental ideas and computational methods."

> Earlier in the paragraph, the authors established their position on the issue, writing, "In our opinion, concepts and computations often positively reinforce one another."

Likewise, Adams et al. (2000, p. 9) critiqued the Singapore materials for their lack of emphasis on higher order thinking skills, writing:

> While the mathematics in Singapore's curriculum may be considered rigorous, we noticed that it does not often engage students in higher order thinking skills. When we examine the types of tasks that the Singapore curriculum asks students to do, we see that Singapore's students are rarely, if ever, asked to analyze, reflect, critique, develop, synthesize, or explain. The vast majority of the student tasks in the Singapore curriculum is based on computation, which primarily reinforces only the recall of facts and procedures. This bias towards certain modes of thinking may be appropriate for an environment in which students' careers depend on the results of a standardized test, but we feel it discourages students from becoming independent learners.

It is possible that at its crux, the debate involves the question of whether higher level skills are best achieved by the careful sequence and accumulation of prerequisite skills or by an early and carefully ordered sequence of cognitive challenges at an appropriate level followed by increasing levels of formalization. Empirical study is needed to address this question. Adjustments in balance may be a long-term outcome of a more carefully articulated set of methods for content analysis.

In summary, the first dimension of content analysis is derived fairly directly from one's knowledge of and perspective on the discipline of mathematics. There is no single view of coherence of mathematics as illustrated by the variations among the examples presented. For some, mathematics curricula derive their coherence from their connection to a set of concepts ordered and sequenced logically, carefully scripted to draw on students' prior knowledge and prepare them for future study, with extensive examples for practice. For others, the coherence is derived from links to applications, opportunities for conjecture, and subsequent developmental progress toward rigor in problem solving and proof, increasing formalization and providing fewer, but more elaborate, examples.

Dimension Two:
Engagement, Timeliness and Support for Diversity, and Assessment

A review of the content analyses reveals very different views of the learner and his/her needs. Clearly, all reviewers claim that their approach best serves the needs of the students, but differences in how this is evaluated emerge among content analyses.

The first criterion we categorized in this dimension is student **engagement**. It was selected to capture a variety of aspects of attention to students' participation in the learning process that may vary because of considerations of prior knowledge, interests, curiosity, compelling misconceptions, alternative perspectives, or motivation. There is a solid research base on many of these issues, and content analysts should establish how they have made use of these data.

The reviews by AAAS provide the strongest examples of using criteria focused on the reviewer's assessment of levels and types of student participation. They analyze the material in terms of its success in providing students with an interesting and compelling purpose, specifying prerequisite knowledge, alerting teachers to student ideas and misconceptions, including a variety of contexts, and providing firsthand experiences. For example, for the Mathematics: Modeling Our World (ARISE) materials, their review states:

> For the three idea sets addressed in the materials [as specified by their methodology], there are an appropriate variety of experiences with objects, applications, and materials that are right on target with the mathematical ideas. They include many tables and graphs, both in the readings and in the exercises; real-world data and content; interesting situations; games; physical activity; and spreadsheets and graphing calculators. Each unit begins with a short video segment and accompanying questions for student response, designed to introduce the unit and provide a context for the mathematics. (Section III.1 Providing Variety of Contexts (2.5)—http://www.project2061.org/tools/textbook/algebra/Mathmode/instrsum/COM_ia3.htm)

For the same materials, the reviewers wrote the following concerning their explicit attention to misconceptions:

> For the Functions Idea Set, the material explicitly addresses commonly held ideas. There are questions, tasks, and activities that are likely to help students progress from their initial ideas by extending correct commonly held ideas that have limited scope. . . .There is no evidence of this kind of

support being provided for the other two idea sets. Although there are some questions and activities related to the Operations Idea Set that are likely to help students progress from initial ideas, students are not asked to compare commonly held ideas with the correct concept. (Section II.4 Addressing Misconceptions (0.8)–http://www.project2061.org/tools/textbook/algebra/Mathmode/instrsum/COM_ia2.htm)

For Saxon, for the same two questions, the AAAS authors wrote:

The experiences provided are mainly pencil and paper activities. The material uses a few different contexts such as calculators and fraction pieces, and in grade six there are three lessons using fraction manipulatives. In grade seven, students put together a flexible model and later work with paper and pencil measurements. A lesson in grade seven on volume of rectangular solids uses a variety of drawings and equations that are right on target with the benchmark, but there is no suggestion that the students actually use sugar cubes, as referenced to build a figure. For the algebra graphs and algebra equation concepts, no variety of contexts is offered. Most firsthand experiences are found in the supplementary materials where students are given a few opportunities to do measurements, work with paper models of figures, collect data, and construct graphs. (Instructional Category III, Engaging Students in Mathematics—http://www.project2061.org/tools/textbook/matheval/13saxon/instruct.htm)

In relation to building on student ideas about mathematics, the same authors wrote:

While there is an occasional reference to prerequisite knowledge, the references are neither consistent nor explicit. In the instances where prerequisite knowledge is identified, the reference is made in the opening narrative that mentions skills that are taught in earlier lessons. The lessons often begin a new skill or procedure without reference to earlier work. There are warm-up activities at the beginning of lessons that provide practice for upcoming skills, but they are not identified as such. There is no guidance for teachers in identifying or addressing student difficulties. (Instructional Category II, Building on Student Ideas about Mathematics—http://www.project2061.org/tools/textbook/matheval/13saxon/instruct.htm)

The next set of passages demonstrates that not all reviewers see the most important source of student engagement as being through the use of context or building on prior knowledge, but rather by the careful choice of student example, sequencing of topic, and adequate and incremental challenge. An example of a content analysis with this focus was found in the reviews of UCSMP by Clopton et al. (1998). The two relevant criteria for review were "the quality and sufficiency of student work" and "range of depth and scope in student work." Summarizing these two criteria under the subtitle of exercises, they wrote:

The number of student exercises is very low, and this is the most blatant negative feature of this text. These exercises are most typically at basic achievement levels with a few moderately difficult problems presented in some instances. For example, the section on solving linear systems by substitution includes six symbolic problems giving very simple systems and three word problems giving very simple systems. The actual number of problems to be solved is less than it appears to be as many of the exercise items are procedure questions. The extent, range, and scope of student work is low enough to cause serious concerns about the consolidation of learning. (Section 3: Overall Evaluation—exercises–http://mathematicallycorrect.com/a1ucsmp.htm)

One can see very different views of engagement in these varied comments, and hence, one would expect varied ratings based on one's meaning for the term.

A persistent criticism found in certain content analyses, but not in others, involves their **timeliness and support for diversity.** We interpret this criterion to apply to meeting the needs of all students, in terms of the level of preparation (high, medium, and low), the diverse perspectives, the cultural resources and backgrounds of students, and the timeliness of the pace of instruction.

As one illustration of the issues subsumed in this criterion, material may be presented so late in the school program that it could jeopardize options for those students going to college or planning a technically oriented career. To support Askey's remark that a "content analysis should consider the structure of the program, whether essential topics have been taught in a timely way," in testimony to the committee, he provided an example where the tardiness in presentation could affect college options. "For Core-Plus, I illustrated how this has not been done by remarking that $(a+b)^2 = a^2 + 2ab + b^2$ is only done in grade 11. This is far too late, for students need time to develop algebra skills, and plenty of problems using algebra to develop the needed skills." Christian Hirsch, Western Michigan University and author of Core-Plus, responded in written testimony, "the topic is treated in CPMP Course 3, pages 212-214 for the beginning of the expansion/factorization development," and that "students study Course 3 in grade 11 or grade 10, if accelerated." Yet including timeliness raises the legitimate issue of whether such a delay in learning this material could put students at a disadvantage when compared with the growing number of students entering college with Advanced Placement calculus and more advanced training in mathematics.

Although absent from some studies, this timeliness theme is consistent through those content analyses that focused on the challenge of the mathematics. To illustrate how this issue can be studied in a content analysis, Adams and colleagues (2000, p. 11) point out, "we find that CMP students are not expected to compute fluently, flexibly and efficiently with fractions, decimals, and percents as late as eighth grade. Standard algorithms for computations with fractions. . . . are often not used. . . . [While] CMP does a good job of helping students discover the mathematical connections and patterns in the algebra strand, [it] falls short in a follow-through with more substantial statements, generalizations, formulas, or algorithms." In an 8th-grade unit, "CMP misses the opportunity to discuss the quadratic formula or the process of completing the square."

Other examples (AAAS, 1999b, Part 1 Conclusions [in box]—http://www.project2061.org/matheval/part1c.htm; Adams et al., 2000, p. 12) could be offered, but the critique permits one to see why the issue of balance is crucial in content analyses, as one examines whether emphasis on discovery approaches and new levels of understanding can carry the cost of a lack of basic knowledge of facts and standard mathematical algorithms at an early age. In addition to timeliness for all students, in many content analyses, there is expressed concern for the most mathematically inclined students to receive enough challenges. For example, Adams et al. (2000, p. 13) wrote that in Mathematics in Context:

> Alongside each lesson are comments about the underlying mathematical concepts in the lesson ("About the Mathematics") as well as how to plan and to actually teach the lesson. A nice feature is that these comments occur in the margins of the Teachers' Guides. . . . On the other hand, these comments often contain some useful mathematical facts and language that could be, but most likely wouldn't be, communicated to the students; in particular high-end students could benefit from these insights if they were available to them. In addition, the lack of a glossary hides mathematical terminology from the students, a language which they should be beginning to negotiate by the middle grades. Exposure to the precise terminology of mathematics is crucial for students at this stage, not only as a means of exemplifying the rigor of mathematics, but as a way to communicate their discoveries and hypotheses in a common language, rather than the idiosyncratic terms that a particular student or class may develop.

As a result of comments such as these, the authors summarize by stating, "high-end students may not find this curriculum very challenging or stimulating" (p. 14).

In the content analyses, support for diversity was typically addressed only in terms of performance levels, and even then, high performers were identified as needing the most attention (Howe testimony). In contrast, other researchers focused on the importance of providing activities that can be used successfully to meet the needs of a variety of student levels of preparation, scaffolding those needing more assistance and including extensions for those ready for more challenge (Schifter testimony). Furthermore, there are other aspects of support for diversity to be considered, such as language use or cultural experiences. More attention in these content analyses must be paid to whether the curriculum serves the diverse needs of students of all ability levels, all language backgrounds, and all cultural roots. Most of the analyses remained at the level of attention to the use of names or pictures portraying occupations by race and gender (AAAS, 1999b; Adams et al., 2000; U.S. Department of Education, 1999). The cognitive dimensions of these students' needs, including remediation, support for reading difficulties, and frequent assessment and review, are less clearly discussed.

Support for diversity in content analyses represents the biggest challenge of all. Scientific approaches have relied mostly on our limited understanding of individual learning and age-dependent cognitive processes. Moreover, efforts to understand these processes have focused at the level of the individual (the "immunology" of learning), while the impact of population forces (i.e., that is, the extrapolation of individual processes at a higher level) on learning is poorly understood (girls, as a group, in 7th and 8th grades are inadequately encouraged to excel in mathematics). Population-level processes can enhance or inhibit learning. These processes may be the biggest obstacle to learning, and curriculum implementations that do not address these forces may fail regardless of the quality discipline-based dimensions of the content analysis, hence the need for learner- and teacher-based dimensions in our framework. The grand challenge is that models that rely solely on traditional scientific approaches may not be successful if the goal is to promote learning in a highly heterogeneous (at many levels) society. Innovative scientific approaches that attend to the big picture and the impact of nonlinear effects at all levels must be adopted.

Within the second dimension (Engagement, Timeliness and Support for Diversity, and Assessment), the final criterion concerns how one determines what students know, or **assessment**. An essential part of examining a curriculum in relation to its effects on students is to examine the various means of assessment. Examining these effects often reveals a great deal about the underlying philosophy of the program.

The quality of attention to assessment in these content analyses is generally weak. In the Mathematically Correct Reviews (Clopton et al., 1998, 1999a, 1999b, and 1999c), assessment is referred to only in terms of "support for student mastery." In the Adams report, which was quite strong in most respects, only two questions are discussed: Does the curriculum include and encourage multiple kinds of assessments (e.g., performance, formative, summative, paper-pencil, observations, portfolios, journals, student interviews, projects)? Does the curriculum provide well-aligned summative assessments to judge a student's attainment? The responses were cursory, such as "This principle is fully met. Well-aligned summative assessments are given at the end of each unit for the teacher's use." The exception to this was found in AAAS's content analyses, where three different components of assessment are reviewed: Are the assessments aligned to the ideas, concepts, and skills of the benchmark? Do they include assessment through applications and not just require repeating memorized terms or rules? Do they use embedded assessments, with advice to teachers on how they might use the results to choose or modify activity? The write-ups for each curriculum show that the reviewers were able to identify important distinctions among curricula along this dimension. For example, in reviewing the assessment practices for each of these three questions concerning the Interactive Mathematics Program (IMP), the reviewers wrote:

> There is at least one assessment item or task that addresses the specific ideas in each of the idea sets, and these items and tasks require no other, more sophisticated ideas. For some idea sets, there are insufficient items that are content-matched to the mathematical ideas. . . . (Section VI.1: Aligning Assessment (2.3)—http://www.project2061.org/tools/textbook/algebra/Interact/instrsum/IMP_ia6.htm)

> All assessment tasks in this material require application of knowledge and/or skills, although ideally would include more applications assessments throughout, and more substantial tasks within these assessments. . . . Assessment tasks alter either the context or the degree of specificity or generalization required, as compared to similar types of problems in class work and homework exercises.

> The material uses embedded assessment as a part of the instructional strategy and design. . . . The authors make a clear distinction between assessment and grading, indicating that assessment occurs throughout each day as teachers gauge the learning process. . . . In spite of all these options for embedded assessment throughout the material, there are few assessments that provide opportunities, encouragement, or guidance for students on how to further understand the mathematical ideas.

AAAS reviews include one additional criterion that is essential to discussions of assessment. It is referred to as "Encouraging students to think about what they have learned" (AAAS, 1999a). The summary of the reviews for IMP is provided to assist in understanding how this criterion adds to one's understanding of the curricular program and suggest places for improvement.

> The material engages students in monitoring their progress toward understanding the mathematical ideas, and only does so primarily through the compilation of a portfolio at the end of each unit. . . . Personal growth is a part of each portfolio as students are encouraged to think about their personal development and how their ideas have developed or changed throughout the unit. However, these reflections are generally very generic, rather than specific to the idea sets. (V.3: Encouraging Students to Think about What They've Learned (1.6)—http://www.project2061.org/tools/textbook/algebra/Interact/instrsum/IMP_ia5.htm)

Dimension Three:
Pedagogy, Resources, and Professional Development

A successful curriculum is impossible if it does not pay attention to the abilities and needs of teachers, and thus **pedagogy** must be a component in a content analysis. Bishop (1997) asserted that "Some elementary school teachers are ill prepared to teach mathematics, and very inexperienced teachers can seriously misjudge how to present the material appropriately to their classes." This concern was often repeated in testimony that we heard from teachers, educators, and textbook publishers. Other researchers saw the curriculum as a vehicle to strengthen teachers' content knowledge and design materials with this purpose in mind (Schifter testimony). The question that arises is how such concerns should affect the conduct of content analyses.

One issue is that such analyses should report on the expectations of the designers for **professional development**. It is clear that programs which introduce new approaches and new technologies will require more professional development for successful implementation. Testimony indicates that even for more traditional curricula, professional development is needed. Deeper understanding is possible only through added support; if stipulated, such requirements should be reported in content analyses. Expanding on this theme, Adams et al. (2000) offer a potential explanation for the poorer U.S. TIMSS performance by commenting that "we must acknowledge that Singapore's educational system—the curriculum, the teachers, the parental support, the social culture, and the strong government support of education—has succeeded in producing students who as a whole understand mathematics at a higher level, and perform with more competence and fluency, than the American students who took the [TIMSS] tests." On the other hand, "[s]imply adopting the middle-grades Singapore curriculum is not likely to help American students move to the top." The issue is far more complex because it also involves teacher development. As the Adams report argued, "The most striking difference between the educational systems of [the United States and Singapore] is in governmental support of education. For example, the government encourages every teacher to attend at least 100 hours of training each year, and has developed an intranet called the 'Teachers' Network,' which enables teachers to share ideas with one another" (Adams et al., 2000, p. A-3). Although this issue of teacher development is addressed in some of the content analyses, it is so crucial that it should be addressed in most of them.

The second criterion in this dimension concerns **resources**, and they, too, must be explicitly considered in content analyses. The Adams report described the important differences between Singapore and the United States in the kind of students being tested and the mathematical experiences they bring to the classroom. As an illustration, the report described how Singapore students are filtered according to talent and how they receive enrichment programs. This information is critical to making an informed comparison, not only about the curriculum itself, but about the assumptions it makes about the environment in which it is to be used. It further helps one to assess the assumptions about resources for curricular use.

Conclusions

In reviewing the 36 content analyses, we determined that although a comprehensive methodology for the conduct of content analyses was lacking, elements were distributed among the submissions. We summarized the content analyses and reported the results when those results could be used inferentially to inform the subsequent conduct of content analyses. We recognized that no amount of delineation of method would produce identical evaluations, but suggested rather that delineation of dimensions for review might help to make content analysis evaluations more informative to curriculum decision makers. Toward this end, we recognized the importance of involvement by mathematicians and mathematics educators in the process, and called for increased participation by practitioners. We discussed the need for careful and thoughtful choices on the standards to be used and the selection of comparative curricula. We acknowledged that some content analysts would focus on the materials a priori, and others would prefer to conduct analysis based on curricular use in situ.

We identified three dimensions along which curricular evaluations of content should be focused:

1. Clarity, comprehensiveness, accuracy, depth of mathematical inquiry and mathematical reasoning, organization, and balance (disciplinary perspectives).
2. Engagement, timeliness and support for diversity, and assessment (learner-oriented perspectives).
3. Pedagogy, resources, and professional development (teacher- and resource-oriented perspectives).

In addition, we recognized that each of these dimensions can be treated differently depending on a reviewer's perspectives of the discipline, the student, and the resources and capacity in schools. A quality content analysis would present a coherent and integrated view of the relationship among these dimensions (Box 4-3).

Differences in content analyses are inevitable and welcome, as they can contribute to providing decision makers with choices among curricula. However, those differences need to be interpreted in relation to an under-

---

BOX 4-3
Characteristics of a Quality Content Analysis of
Mathematics Curriculum Materials

- Examines the **clarity of objectives** and their **comprehensiveness** of treatment of identified standards in relation to specified comparative curricula.
- Determines the **accuracy** and **depth of mathematical inquiry and mathematical reasoning** in the curriculum materials.
- Evaluates the **balance** of curricular choices such as conceptual/procedural; use of context versus decontextualized treatments; informal versus formal; and varied uses of representational forms.
- Examines the **engagement** of students.
- Discusses the **timeliness and support for diversity** of the curriculum materials in relation to the particular grade level(s) for which the materials are designated.
- Discusses **assessment** of what is learned.
- Discusses **teacher capacity and training, resources, and professional development** needed to present the curriculum materials.

---

lying set of dimensions for comparison. Therefore, we have provided an articulation of the underlying dimensions that could lead to a kind of consumer's guide to assist the readers of multiple sources of content analysis. One would expect that just as one develops preferences in reviewers (whether they are reviewing books, wine, works of art, or movies), one will select one's advisers on the basis of preferences and preferred expertise and perspectives. Second, with a more explicit discussion of methodology, reviewers with divergent perspectives may find more effective means of identifying and discussing differences. This might involve the use of panels of reviewers in discussion of particular examples with contrasting perspectives.

In these content analyses—over time and particularly in the solicited letters—we see evidence that the polarization that characterizes the mathematics education communities (mathematicians, mathematics educators, teachers, parents) could be partially reconciled by more mutual acknowledgment of the legitimacy of diverse perspectives and the shared practical need to serve children better. It is possible, for example, that based on critical content analyses, subsequent versions of reform curricula could be revised to strengthen weak or incomplete areas, traditional curricular materials could be revised to provide more uses of innovative methods, and new hybrids of the approaches could be developed.

At the same time, while the issue of content analysis is critical from philosophical, academic, and logical viewpoints, it is not clear what degree or type of impact current content changes have on student learning and achievement when curricula are implemented in classrooms. Learning is taking place in a complex dynamical system, and one should consider whether the correction of current curricular deficiencies, a process that obviously must be carried out, is the key pressure point for changing the educational system. Will curricular content changes by themselves actually reduce the disparities in learning that have been documented in many studies? Answers to analogous questions exist in other fields. For example, although detailed understanding of the immunology of HIV (resulting in the development of a robust vaccine) may be the key in the long run, as of today, human behavior along with cultural practices and norms are the drivers (pressure points) of the HIV epidemic. Efforts to alter behavior and cultural practices and norms, at this point, are most likely to have immediate impact on HIV transmission dynamics. Furthermore, changes in the social landscape will have a beneficial impact on the transmission dynamics and control of many sexually transmitted diseases. Analogously, in examining curricular effectiveness the role of content analyses, while critical, is not the only pressure point in the system and may interact in fundamental ways with other pressure points. Therefore it is essential to consider methods and approaches that take into account the possible impact and implications of changing and evolving landscapes on a curriculum's implementation and effectiveness. The landscape shift illustrated in the 5-year longitudinal study by Carroll (2001) of Everyday Mathematics highlights some of these challenges. The next two chapters, on comparative studies and case studies, add indirect consideration of other influences on the curricular effectiveness.

Nonetheless, as we transition to these, we emphasize the importance of content analysis in relation to comparative and case studies. For example, content analyses can offer insights into the designs of comparative analyses. A thorough content analysis provides a clear articulation of the program theory from one point of view. From a set of such reviews, researchers can identify contentious issues that merit further basic research. One could examine questions such as, Do clear, concise materials used as primary texts lead to stronger conceptual development than engaging, active challenges and tasks? or Does the introduction of formal definitions facilitate further and deeper understanding and mastery of new material? How and when does this work? Research is needed to determine whether analyses of the intended curricula are validated by the empirical outcomes of the enacted curricula. Content analyses are also valuable to inform the conduct of comparative studies. A content analysis can help an evaluator to select appropriate outcome measures, to measure particularly important content strands, and to concentrate on the essential aspects of implementation and professional development. For these multiple reasons, careful and increasingly sophisticated content analysis will make important contributions to the evaluation of the effectiveness of a curricular program.

5

Comparative Studies

It is deceptively simple to imagine that a curriculum's effectiveness could be easily determined by a single well-designed study. Such a study would randomly assign students to two treatment groups, one using the experimental materials and the other using a widely established comparative program. The students would be taught the entire curriculum, and a test administered at the end of instruction would provide unequivocal results that would permit one to identify the more effective treatment.

The truth is that conducting definitive comparative studies is not simple, and many factors make such an approach difficult. Student placement and curricular choice are decisions that involve multiple groups of decision makers, accrue over time, and are subject to day-to-day conditions of instability, including student mobility, parent preference, teacher assignment, administrator and school board decisions, and the impact of standardized testing. This complex set of institutional policies, school contexts, and individual personalities makes comparative studies, even quasi-experimental approaches, challenging, and thus demands an honest and feasible assessment of what can be expected of evaluation studies (Usiskin, 1997; Kilpatrick, 2002; Schoenfeld, 2002; Shafer, in press).

Comparative evaluation study is an evolving methodology, and our purpose in conducting this review was to evaluate and learn from the efforts undertaken so far and advise on future efforts. We stipulated the use of comparative studies as follows:

A comparative study was defined as a study in which two (or more) curricular treatments were investigated over a substantial period of time (at least one semester, and more typically an entire school year) and a comparison of various curricular outcomes was examined using statistical tests. A statistical test was required to ensure the robustness of the results relative to the study's design.

We read and reviewed a set of 95 comparative studies. In this report we describe that database, analyze its results, and draw conclusions about the quality of the evaluation database both as a whole and separated into evaluations supported by the National Science Foundation and commercially generated evaluations. In addition to describing and analyzing this database, we also provide advice to those who might wish to fund or conduct future comparative evaluations of mathematics curricular effectiveness. We have concluded that the process of conducting such evaluations is in its adolescence and could benefit from careful synthesis and advice in order to increase its rigor, feasibility, and credibility. In addition, we took an interdisciplinary approach to the task, noting that various committee members brought different expertise and priorities to the consideration of what constitutes the most essential qualities of rigorous and valid experimental or quasi-experimental design in evaluation. This interdisciplinary approach has led to some interesting observations and innovations in our methodology of evaluation study review.

This chapter is organized as follows:

- Study counts disaggregated by program and program type.
- Seven critical decision points and identification of *at least minimally methodologically adequate* studies.
- Definition and illustration of each decision point.
- A summary of results by student achievement in relation to program types (NSF-supported, University of Chicago School Mathematics Project (UCSMP), and commercially generated) in relation to their reported outcome measures.
- A list of alternative hypotheses on effectiveness.
- Filters based on the critical decision points.
- An analysis of results by subpopulations.
- An analysis of results by content strand.
- An analysis of interactions among content, equity, and grade levels.
- Discussion and summary statements.

In this report, we describe our methodology for review and synthesis so that others might scrutinize our approach and offer criticism on the basis of our methodology and its connection to the results stated and conclusions drawn. In the spirit of scientific, fair, and open investigation, we welcome others to undertake similar or contrasting approaches and compare and discuss the results. Our work was limited by the short timeline set by the funding agencies resulting from the urgency of the task. Although we made multiple efforts to collect comparative studies, we apologize to any curriculum evaluators if comparative studies were unintentionally omitted from our database.

Of these 95 comparative studies, 65 were studies of NSF-supported curricula, 27 were studies of commercially generated materials, and 3 included two curricula each from one of these two categories. To avoid the problem of double coding, two studies, White et al. (1995) and Zahrt (2001), were coded within studies of NSF-supported curricula because more of the classes studied used the NSF-supported curriculum. These studies were not used in later analyses because they did not meet the requirements for the *at least minimally methodologically adequate* studies, as described below. The other, Peters (1992), compared two commercially generated curricula, and was coded in that category under the primary program of focus. Therefore, of the 95 comparative studies, 67 studies were coded as NSF-supported curricula and 28 were coded as commercially generated materials.

The 11 evaluation studies of the UCSMP secondary program that we reviewed, not including White et al. and Zahrt as previously mentioned, benefit from the maturity of the program, while demonstrating an orientation to both establishing effectiveness and improving a product line. For these reasons, at times we will present the summary of UCSMP's data separately.

The Saxon materials also present a somewhat different profile from the other commercially generated materials because many of the evaluations of these materials were conducted in the 1980s and the materials were originally developed with a rather atypical program theory. Saxon (1981) designed its algebra materials to combine distributed practice with incremental development. We selected the Saxon materials as a middle grades commercially generated program, and limited its review to middle school studies from 1989 onward when the first National Council of Teachers of Mathematics (NCTM) Standards (NCTM, 1989) were released. This eliminated concerns that the materials or the conditions of educational practice have been altered during the intervening time period. The Saxon materials explicitly do not draw from the NCTM Standards nor did they receive support from the NSF; thus they truly represent a commercial venture. As a result, we categorized the Saxon studies within the group of studies of commercial materials.

At times in this report, we describe characteristics of the database by

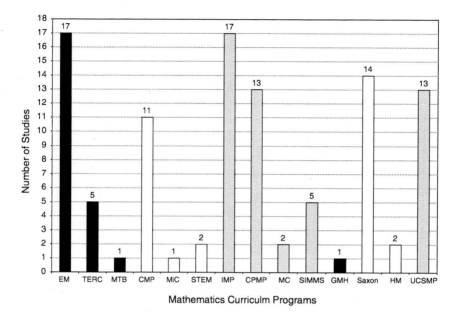

FIGURE 5-1 The distribution of comparative studies across programs.
Programs are coded by grade band: black bars = elementary, white bars = middle grades, and gray bars = secondary. In this figure, there are six studies that involved two programs and one study that involved three programs.
NOTE: Five programs (MathScape, MMAP, MMOW/ARISE, Addison-Wesley, and Harcourt) are not shown above since no comparative studies were reviewed.

particular curricular program evaluations, in which case all 19 programs are listed separately. At other times, when we seek to inform ourselves on policy-related issues of funding and evaluating curricular materials, we use the NSF-supported, commercially generated, and UCSMP distinctions. We remind the reader of the artificial aspects of this distinction because at the present time, 18 of the 19 curricula are published commercially. In order to track the question of historical inception and policy implications, a distinction is drawn between the three categories. Figure 5-1 shows the distribution of comparative studies across the 14 programs.

The first result the committee wishes to report is the uneven distribution of studies across the curricula programs. There were 67 coded studies of the NSF curricula, 11 studies of UCSMP, and 17 studies of the commercial publishers. The 14 evaluation studies conducted on the Saxon materials compose the bulk of these 17-non-UCSMP and non-NSF-supported curricular evaluation studies. As these results suggest, we know more about the evaluations of the NSF-supported curricula and UCSMP than about the evaluations of the commercial programs. We suggest that three factors account for this uneven distribution of studies. First, evaluations have been funded by the NSF both as a part of the original call, and as follow-up to the work in the case of three supplemental awards to two of the curricula programs. Second, most NSF-supported programs and UCSMP were developed at university sites where there is access to the resources of graduate students and research staff. Finally, there was some reported reluctance on the part of commercial companies to release studies that could affect perceptions of competitive advantage. As Figure 5-1 shows, there were quite a few comparative studies of Everyday Mathematics (EM), Connected Mathematics Project (CMP), Contemporary Mathematics in Context (Core-Plus Mathematics Project [CPMP]), Interactive Mathematics Program (IMP), UCSMP, and Saxon.

In the programs with many studies, we note that a significant number of studies were generated by a core set of authors. In some cases, the evaluation reports follow a relatively uniform structure applied to single schools, generating multiple studies or following cohorts over years. Others use a standardized evaluation approach to evaluate sequential courses. Any reports duplicating exactly the same sample, outcome measures, or forms of analysis were eliminated. For example, one study of Mathematics Trailblazers (Carter et al., 2002) reanalyzed the data from the larger ARC Implementation Center study (Sconiers et al., 2002), so it was not included separately. Synthesis studies referencing a variety of evaluation reports are summarized in Chapter 6, but relevant individual studies that were referenced in them were sought out and included in this comparative review.

Other less formal comparative studies are conducted regularly at the school or district level, but such studies were not included in this review unless we could obtain formal reports of their results, and the studies met the criteria outlined for inclusion in our database. In our conclusions, we address the issue of how to collect such data more systematically at the district or state level in order to subject the data to the standards of scholarly peer review and make it more systematically and fairly a part of the national database on curricular effectiveness.

A standard for evaluation of any social program requires that an impact assessment is warranted only if two conditions are met: (1) the curricular program is clearly specified, and (2) the intervention is well implemented. Absent this assurance, one must have a means of ensuring or measuring treatment integrity in order to make causal inferences. Rossi et al. (1999, p. 238) warned that:

> two prerequisites [must exist] for assessing the impact of an intervention.
> First, the program's objectives must be sufficiently well articulated to make

it possible to specify credible measures of the expected outcomes, or the evaluator must be able to establish such a set of measurable outcomes. Second, the intervention should be sufficiently well implemented that there is no question that its critical elements have been delivered to appropriate targets. It would be a waste of time, effort, and resources to attempt to estimate the impact of a program that lacks measurable outcomes or that has not been properly implemented. An important implication of this last consideration is that interventions should be evaluated for impact only when they have been in place long enough to have ironed out implementation problems.

These same conditions apply to evaluation of mathematics curricula. The comparative studies in this report varied in the quality of documentation of these two conditions; however, all addressed them to some degree or another. Initially by reviewing the studies, we were able to identify one general design template, which consisted of seven critical decision points and determined that it could be used to develop a framework for conducting our meta-analysis. The seven critical decision points we identified initially were:

1. Choice of type of design: experimental or quasi-experimental;
2. For those studies that do not use random assignment: what methods of establishing comparability of groups were built into the design—this includes student characteristics, teacher characteristics, and the extent to which professional development was involved as part of the definition of a curriculum;
3. Definition of the appropriate unit of analysis (students, classes, teachers, schools, or districts);
4. Inclusion of an examination of implementation components;
5. Definition of the outcome measures and disaggregated results by program;
6. The choice of statistical tests, including statistical significance levels and effect size; and
7. Recognition of limitations to generalizability resulting from design choices.

These are critical decisions that affect the quality of an evaluation. We further identified a subset of these evaluation studies that met a set of minimum conditions that we termed *at least minimally methodologically adequate* studies. Such studies are those with the greatest likelihood of shedding light on the effectiveness of these programs. To be classified as *at least minimally methodologically adequate,* and therefore to be considered for further analysis, each evaluation study was required to:

• Include quantifiably measurable outcomes such as test scores, responses to specified cognitive tasks of mathematical reasoning, performance evaluations, grades, and subsequent course taking; and
• Provide adequate information to judge the comparability of samples.

In addition, a study must have included at least one of the following additional design elements:

• A report of implementation fidelity or professional development activity;
• Results disaggregated by content strands or by performance by student subgroups; and/or
• Multiple outcome measures or precise theoretical analysis of a measured construct, such as number sense, proof, or proportional reasoning.

Using this rubric, the committee identified a subset of 63 comparative studies to classify as *at least minimally methodologically adequate* and to analyze in depth to inform the conduct of future evaluations. There are those who would argue that any threat to the validity of a study discredits the findings, thus claiming that until we know everything, we know nothing. Others would claim that from the myriad of studies, examining patterns of effects and patterns of variation, one can learn a great deal, perhaps tentatively, about programs and their possible effects. More importantly, we can learn about methodologies and how to concentrate and focus to increase the likelihood of learning more quickly. As Lipsey (1997, p. 22) wrote:

> In the long run, our most useful and informative contribution to program managers and policy makers and even to the evaluation profession itself may be the consolidation of our piecemeal knowledge into broader pictures of the program and policy spaces at issue, rather than individual studies of particular programs.

We do not wish to imply that we devalue studies of student affect or conceptions of mathematics, but decided that unless these indicators were connected to direct indicators of student learning, we would eliminate them from further study. As a result of this sorting, we eliminated 19 studies of NSF-supported curricula and 13 studies of commercially generated curricula. Of these, 4 were eliminated for their sole focus on affect or conceptions, 3 were eliminated for their comparative focus on outcomes other than achievement, such as teacher-related variables, and 19 were eliminated for their failure to meet the minimum additional characteristics specified in the criteria above. In addition, six others were excluded from the studies of commercial materials because they were not conducted within the grade-level band specified by the committee for the selection of that program. From this point onward, all references can be assumed to refer to *at least minimally methodologically adequate* unless a study is referenced for illustration, in which case we label it with "EX" to indicate that it is excluded in the summary analyses. Studies labeled "EX" are occasionally referenced because they can provide useful information on certain aspects of curricular evaluation, but not on the overall effectiveness.

The *at least minimally methodologically adequate* studies reported on a variety of grade levels. Figure 5-2 shows the different grade levels of the studies. At times, the choice of grade levels was dictated by the years in which high-stakes tests were given. Most of the studies reported on multiple grade levels, as shown in Figure 5-2.

Using the seven critical design elements of *at least minimally methodologically adequate* studies as a design template, we describe the overall database and discuss the array of choices on critical decision points with examples. Following that, we report on the results on the *at least minimally methodologically adequate* studies by program type. To do so, the results of each study were coded as either statistically significant or not. Those studies

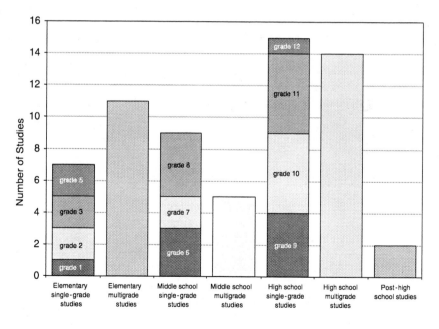

FIGURE 5-2 Single-grade studies by grade and multigrade studies by grade band.

that contained statistically significant results were assigned a percentage of outcomes that are positive (in favor of the treatment curriculum) based on the number of statistically significant comparisons reported relative to the total number of comparisons reported, and a percentage of outcomes that are negative (in favor of the comparative curriculum). The remaining were coded as the percentage of outcomes that are non significant. Then, using seven critical decision points as filters, we identified and examined more closely sets of studies that exhibited the strongest designs, and would therefore be most likely to increase our confidence in the validity of the evaluation. In this last section, we consider alternative hypotheses that could explain the results.

The committee emphasizes that we did not directly evaluate the materials. We present no analysis of results aggregated across studies by naming individual curricular programs because we did not consider the magnitude or rigor of the database for individual programs substantial enough to do so. Nevertheless, there are studies that provide compelling data concerning the effectiveness of the program in a particular context. Furthermore, we do report on individual *studies* and their results to highlight issues of approach and methodology and to remain within our primary charge, which was to evaluate the evaluations, we do not summarize results of the individual programs.

DESCRIPTION OF COMPARATIVE STUDIES DATABASE ON CRITICAL DECISION POINTS

An Experimental or Quasi-Experimental Design

We separated the studies into experimental and quasiexperimental, and found that 100 percent of the studies were quasiexperimental (Campbell and Stanley, 1966; Cook and Campbell, 1979; and Rossi et al., 1999).[1] Within the quasi-experimental studies, we identified three subcategories of comparative study. In the first case, we identified a study as cross-curricular comparative if it compared the results of curriculum A with curriculum B. A few studies in this category also compared two samples within the curriculum to each other and specified different conditions such as high and low implementation quality.

A second category of a quasi-experimental study involved comparisons that could shed light on effectiveness involving time series studies. These studies compared the performance of a sample of students in a curriculum

---

[1]One study, by Peters (1992), used random assignment to two classrooms, but was classified as quasi-experimental with its sample size and use of qualitative methods.

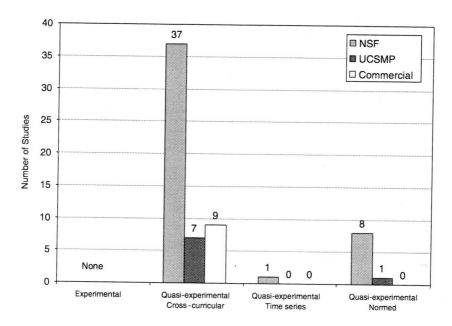

FIGURE 5-3 The number of comparative studies in each category.

under investigation across time, such as in a longitudinal study of the same students over time. A third category of comparative study involved a comparison to some form of externally normed results, such as populations taking state, national, or international tests or prior research assessment from a published study or studies. We categorized these studies and divided them into NSF, UCSMP, and commercial and labeled them by the categories above (Figure 5-3).

In nearly all studies in the comparative group, the titles of experimental curricula were explicitly identified. The only exception to this was the ARC Implementation Center study (Sconiers et al., 2002), where three NSF-supported elementary curricula were examined, but in the results, their effects were pooled. In contrast, in the majority of the cases, the comparison curriculum is referred to simply as "traditional." In only 22 cases were comparisons made between two identified curricula. Many others surveyed the array of curricula at comparison schools and reported on the most frequently used, but did not identify a single curriculum. This design strategy is used often because other factors were used in selecting comparison groups, and the additional requirement of a single identified curriculum in these sites would often make it difficult to match. Studies were categorized into specified (including a single or multiple identified curricula) and nonspecified curricula. In the 63 studies, the central group was compared to an NSF-supported curriculum (1), an unnamed traditional curriculum (41), a named traditional curriculum (19), and one of the six commercial curricula (2). To our knowledge, any systematic impact of such a decision on results has not been studied, but we express concern that when a specified curriculum is compared to an unspecified content which is a set of many informal curriculum, the comparison may favor the coherency and consistency of the single curricula, and we consider this possibility subsequently under alternative hypotheses. We believe that a quality study should at least report the array of curricula that comprise the comparative group and include a measure of the frequency of use of each, but a well-defined alternative is more desirable.

If a study was both longitudinal and comparative, then it was coded as comparative. When studies only examined performances of a group over time, such as in some longitudinal studies, it was coded as quasi-experimental normed. In longitudinal studies, the problems created by student mobility were evident. In one study, Carroll (2001), a five-year longitudinal study of Everyday Mathematics, the sample size began with 500 students, 24 classrooms, and 11 schools. By 2nd grade, the longitudinal sample was 343. By 3rd grade, the number of classes increased to 29 while the number of original students decreased to 236 students. At the completion of the study, approximately 170 of the original students were still in the sample. This high rate of attrition from the study suggests that mobility is a major challenge in curricular evaluation, and that the effects of curricular change on mobile students needs to be studied as a potential threat to the validity of the comparison. It is also a challenge in curriculum implementation because students coming into a program do not experience its cumulative, developmental effect.

Longitudinal studies also have unique challenges associated with outcome measures, a study by Romberg et al. (in press) (EX) discussed one approach to this problem. In this study, an external assessment system and a problem-solving assessment system were used. In the External Assessment System, items from the National Assessment of Educational Progress (NAEP) and Third International Mathematics and Science Survey (TIMSS) were balanced across four strands (number, geometry, algebra, probability and statistics), and 20 items of moderate difficulty, called anchor items, were repeated on each grade-specific assessment (p. 8). Because the analyses of the results are currently under way, the evaluators could not provide us with final results of this study, so it is coded as EX.

However, such longitudinal studies can provide substantial evidence of the effects of a curricular program because they may be more sensitive to an

TABLE 5-1 Scores in Percentage Correct by Everyday Mathematics Students and Various Comparison Groups Over a Five-Year Longitudinal Study

	Sample Size	1st Grade	2nd Grade	3rd Grade	4th Grade	5th Grade
EM	n=170-503	58	62	61	71	75
Traditional U.S.	n=976	43	53.5			44
Japanese	n=750	64	71			80
Chinese	n=1,037	52				76
NAEP Sample	n=18,033				44	44

NOTE: 1st grade: 44 items; 2nd grade: 24 items; 3rd grade: 22 items; 4th grade: 29 items; and 5th grade: 33 items.
SOURCE: Adapted from Carroll (2001).

accumulation of modest effects and/or can reveal whether the rates of learning change over time within curricular change.

The longitudinal study by Carroll (2001) showed that the effects of curricula may often accrue over time, but measurements of achievement present challenges to drawing such conclusions as the content and grade level change. A variety of measures were used over time to demonstrate growth in relation to comparison groups. The author chose a set of measures used previously in studies involving two Asian samples and an American sample to provide a contrast to the students in EM over time. For 3rd and 4th grades, where the data from the comparison group were not available, the authors selected items from the NAEP to bridge the gap. Table 5-1 summarizes the scores of the different comparative groups over five years. Scores are reported as the mean percentage correct for a series of tests on number computation, number concepts and applications, geometry, measurement, and data analysis.

It is difficult to compare performances on different tests over different groups over time against a single longitudinal group from EM, and it is not possible to determine whether the students' performance is increasing or whether the changes in the tests at each grade level are producing the results; thus the results from longitudinal studies lacking a control group or use of sophisticated methodological analysis may be suspect and should be interpreted with caution.

In the Hirsch and Schoen (2002) study, based on a sample of 1,457 students, scores on Ability to Do Quantitative Thinking (ITED-Q) a subset of the Iowa Tests of Education Development, students in Core-Plus showed increasing performance over national norms over the three-year time period. The authors describe the content of the ITED-Q test and point out that "although very little symbolic algebra is required, the ITED-Q is quite demanding for the full range of high school students" (p. 3). They further point out that "[t]his 3-year pattern is consistent, on average, in rural, urban, and suburban schools, for males and females, for various minority groups, and for students for whom English was not their first language" (p. 4). In this case, one sees that studies over time are important as results over shorter periods may mask cumulative effects of consistent and coherent treatments and such studies could also show increases that do not persist when subject to longer trajectories. One approach to longitudinal studies was used by Webb and Dowling in their studies of the Interactive Mathematics Program (Webb and Dowling, 1995a, 1995b, 1995c). These researchers conducted transcript analyses as a means to examine student persistence and success in subsequent course taking.

The third category of quasi-experimental comparative studies measured student outcomes on a particular curricular program and simply compared them to performance on national tests or international tests. When these tests were of good quality and were representative of a genuine sample of a relevant population, such as NAEP reports or TIMSS results, the reports often provided one a reasonable indicator of the effects of the program if combined with a careful description of the sample. Also, sometimes the national tests or state tests used were norm-referenced tests producing national percentiles or grade-level equivalents. The normed studies were considered of weaker quality in establishing effectiveness, but were still considered valid as examples of comparing samples to populations.

For Studies That Do Not Use Random Assignment: What Methods of Establishing Comparability Across Groups Were Built into the Design

The most fundamental question in an evaluation study is whether the treatment has had an effect on the chosen criterion variable. In our context, the treatment is the curriculum materials, and in some cases, related professional development, and the outcome of interest is academic learning. To establish if there is a treatment effect, one must logically rule out as many other explanations as possible for the differences in the outcome variable. There is a long tradition on how this is best done, and the principle from a design point of view is to assure that there are no differences between the treatment conditions (especially in these evaluations, often there are only the new curriculum materials to be evaluated and a control group) either at the outset of the study or during the conduct of the study.

To ensure the first condition, the ideal procedure is the random assignment of the appropriate units to the treatment conditions. The second condition requires that the treatment is administered reliably during the length of the study, and is assured through the careful observation and control of the situation. Without randomization, there are a host of possible confounding variables that could differ among the treatment conditions and that are related themselves to the outcome variables. Put another way, the treatment effect is a parameter that the study is set up to estimate. Statistically, an estimate that is unbiased is desired. The goal is that its expected value over repeated samplings is equal to the true value of the parameter. Without randomization at the onset of a study, there is no way to assure this property of unbiasness. The variables that differ across treatment conditions and are related to the outcomes are confounding variables, which bias the estimation process.

Only one study we reviewed, Peters (1992), used randomization in the assignment of students to treatments, but that occurred because the study was limited to one teacher teaching two sections and included substantial qualitative methods, so we coded it as quasi-experimental. Others report partially assigning teachers randomly to treatment conditions (Thompson, et al., 2001; Thompson et al., 2003). Two primary reasons seem to account for a lack of use of pure experimental design. To justify the conduct and expense of a randomized field trial, the program must be described adequately and there must be relative assurance that its implementation has occurred over the duration of the experiment (Peterson et al., 1999). Additionally, one must be sure that the outcome measures are appropriate for the range of performances in the groups and valid relative to the curricula under investigation. Seldom can such conditions be assured for all students and teachers and over the duration of a year or more.

A second reason is that random assignment of classrooms to curricular treatment groups typically is not permitted or encouraged under normal school conditions. As one evaluator wrote, "Building or district administrators typically identified teachers who would be in the study and in only a few cases was random assignment of teachers to UCSMP Algebra or comparison classes possible. School scheduling and teacher preference were more important factors to administrators and at the risk of losing potential sites, we did not insist on randomization" (Mathison et al., 1989, p. 11).

The Joint Committee on Standards for Educational Evaluation (1994, p. 165) committee of evaluations recognized the likelihood of limitations on randomization, writing:

> The groups being compared are seldom formed by random assignment. Rather, they tend to be natural groupings that are likely to differ in various ways. Analytical methods may be used to adjust for these initial differences, but these methods are based upon a number of assumptions. As it is often difficult to check such assumptions, it is advisable, when time and resources permit, to use several different methods of analysis to determine whether a replicable pattern of results is obtained.

Does the dearth of pure experimentation render the results of the studies reviewed worthless? Bias is not an "either-or" proposition, but it is a quantity of varying degrees. Through careful measurement of the most salient potential confounding variables, precise theoretical description of constructs, and use of these methods of statistical analysis, it is possible to reduce the amount of bias in the estimated treatment effect. Identification of the most likely confounding variables and their measurement and subsequent adjustments can greatly reduce bias and help estimate an effect that is likely to be more reflective of the true value. The theoretical fully specified model is an alternative to randomization by including relevant variables and thus allowing the unbiased estimation of the parameter. The only problem is realizing when the model is fully specified.

We recognized that we can never have enough knowledge to assure a fully specified model, especially in the complex and unstable conditions of schools. However, a key issue in determining the degree of confidence we have in these evaluations is to examine how they have identified, measured, or controlled for such confounding variables. In the next sections, we report on the methods of the evaluators in identifying and adjusting for such potential confounding variables.

One method to eliminate confounding variables is to examine the extent to which the samples investigated are equated either by sample selection or by methods of statistical adjustments. For individual students, there is a large literature suggesting the importance of social class to achievement. In addition, prior achievement of students must be considered. In the comparative studies, investigators first identified participation of districts, schools, or classes that could provide sufficient duration of use of curricular materials (typically two years or more), availability of target classes, or adequate levels of use of program materials. Establishing comparability was a secondary concern.

These two major factors were generally used in establishing the comparability of the sample:

1. Student population characteristics, such as demographic characteristics of students in terms of race/ethnicity, economic levels, or location type (urban, suburban, or rural).
2. Performance-level characteristics such as performance on prior tests, pretest performance, percentage passing standardized tests, or related measures (e.g., problem solving, reading).

In general, four methods of comparing groups were used in the studies we examined, and they permit different degrees of confidence in their results. In the first type, a matching class, school, or district was identified.

Studies were coded as this type if specified characteristics were used to select the schools systematically. In some of these studies, the methodology was relatively complex as correlates of performance on the outcome measures were found empirically and matches were created on that basis (Schneider, 2000; Riordan and Noyce, 2001; and Sconiers et al., 2002). For example, in the Sconiers et al. study, where the total sample of more than 100,000 students was drawn from five states and three elementary curricula are reviewed (Everyday Mathematics, Math Trailblazers [MT], and Investigations [IN], a highly systematic method was developed. After defining eligibility as a "reform school," evaluators conducted separate regression analyses for the five states at each tested grade level to identify the strongest predictors of average school mathematics score. They reported, "reading score and low-income variables . . . consistently accounted for the greatest percentage of total variance. These variables were given the greatest weight in the matching process. Other variables—such as percent white, school mobility rate, and percent with limited English proficiency (LEP)—accounted for little of the total variance but were typically significant. These variables were given less weight in the matching process" (Sconiers et al., 2002, p. 10). To further provide a fair and complete comparison, adjustments were made based on regression analysis of the scores to minimize bias prior to calculating the difference in scores and reporting effect sizes. In their results the evaluators report, "The combined state-grade effect sizes for math and total are virtually identical and correspond to a percentile change of about 4 percent favoring the reform students" (p. 12).

A second type of matching procedure was used in the UCSMP evaluations. For example, in an evaluation centered on geometry learning, evaluators advertised in NCTM and UCSMP publications, and set conditions for participation from schools using their program in terms of length of use and grade level. After selecting schools with heterogeneous grouping and no tracking, the researchers used a match-pair design where they selected classes from the same school on the basis of mathematics ability. They used a pretest to determine this, and because the pretest consisted of two parts, they adjusted their significance level using the Bonferroni method.[2] Pairs were discarded if the differences in means and variance were significant for all students or for those students completing all measures, or if class sizes became too variable. In the algebra study, there were 20 pairs as a result of the matching, and because they were comparing three experimental conditions—first edition, second edition, and comparison classes—in the com-

---

[2]The Bonferroni method is a simple method that allows multiple comparison statements to be made (or confidence intervals to be constructed) while still assuring that an overall confidence coefficient is maintained.

parison study relevant to this review, their matching procedure identified 8 pairs. When possible, teachers were assigned randomly to treatment conditions. Most results are presented with the eight identified pairs and an accumulated set of means. The outcomes of this particular study are described below in a discussion of outcome measures (Thompson et al., 2003).

A third method was to measure factors such as prior performance or socio-economic status (SES) based on pretesting, and then to use analysis of covariance or multiple regression in the subsequent analysis to factor in the variance associated with these factors. These studies were coded as "control." A number of studies of the Saxon curricula used this method. For example, Rentschler (1995) conducted a study of Saxon 76 compared to Silver Burdett with 7th graders in West Virginia. He reported that the groups differed significantly in that the control classes had 65 percent of the students on free and reduced-price lunch programs compared to 55 percent in the experimental conditions. He used scores on California Test of Basic Skills mathematics computation and mathematics concepts and applications as his pretest scores and found significant differences in favor of the experimental group. His posttest scores showed the Saxon experimental group outperformed the control group on both computation and concepts and applications. Using analysis of covariance, the computation difference in favor of the experimental group was statistically significant; however, the difference in concepts and applications was adjusted to show no significant difference at the $p < .05$ level.

A fourth method was noted in studies that used less rigorous methods of selection of sample and comparison of prior achievement or similar demographics. These studies were coded as "compare." Typically, there was no explicit procedure to decide if the comparison was good enough. In some of the studies, it appeared that the comparison was not used as a means of selection, but rather as a more informal device to convince the reader of the plausibility of the equivalence of the groups. Clearly, the studies that used a more precise method of selection were more likely to produce results on which one's confidence in the conclusions is greater.

Definition of Unit of Analysis

A major decision in forming an evaluation design is the unit of analysis. The unit of selection or randomization used to assign elements to treatment and control groups is closely linked to the unit of analysis. As noted in the National Research Council (NRC) report (1992, p. 21):

> If one carries out the assignment of treatments at the level of schools, then that is the level that can be justified for causal analysis. To analyze the results at the student level is to introduce a new, nonrandomized level into

the study, and it raises the same issues as does the nonrandomized observational study. . . . The implications . . . are twofold. First, it is advisable to use randomization at the level at which units are most naturally manipulated. Second, when the unit of observation is at a "lower" level of aggregation than the unit of randomization, then for many purposes the data need to be aggregated in some appropriate fashion to provide a measure that can be analyzed at the level of assignment. Such aggregation may be as simple as a summary statistic or as complex as a context-specific model for association among lower-level observations.

In many studies, inadequate attention was paid to the fact that the unit of selection would later become the unit of analysis. The unit of analysis, for most curriculum evaluators, needs to be at least the classroom, if not the school or even the district. The units must be independently responding units because instruction is a group process. Students are not independent, the classroom—even if the teachers work together in a school on instruction—is not entirely independent, so the school is the unit. Care needed to be taken to ensure that an adequate numbers of units would be available to have sufficient statistical power to detect important differences.

A curriculum is experienced by students in a group, and this implies that individual student responses and what they learn are correlated. As a result, the appropriate unit of assignment and analysis must at least be defined at the classroom or teacher level. Other researchers (Bryk et al., 1993) suggest that the unit might be better selected at an even higher level of aggregation. The school itself provides a culture in which the curriculum is enacted as it is influenced by the policies and assignments of the principal, by the professional interactions and governance exhibited by the teachers as a group, and by the community in which the school resides. This would imply that the school might be the appropriate unit of analysis. Even further, to the extent that such decisions about curriculum are made at the district level and supported through resources and professional development at that level, the appropriate unit could arguably be the district. On a more practical level, we found that arguments can be made for a variety of decisions on the selection of units, and what is most essential is to make a clear argument for one's choice, to use the same unit in the analysis as in the sample selection process, and to recognize the potential limits to generalization that result from one's decisions.

We would argue in all cases that reports of how sites are selected must be explicit in the evaluation report. For example, one set of evaluation studies selected sites by advertisements in a journal distributed by the program and in NCTM journals (UCSMP) (Thompson et al., 2001; Thompson et al., 2003). The samples in their studies tended to be affluent suburban populations and predominantly white populations. Other conditions of inclusion, such as frequency of use also might have influenced this outcome, but it is important that over a set of studies on effectiveness, all populations of students be adequately sampled. When a study is not randomized, adjustments for these confounding variables should be included. In our analysis of equity, we report on the concerns about representativeness of the overall samples and their impact on the generalizability of the results.

Implementation Components

The complexity of doing research on curricular materials introduces a number of possible confounding variables. Due to the documented complexity of curricular implementation, most comparative study evaluators attempt to monitor implementation in some fashion. A valuable outcome of a well-conducted evaluation is to determine not only if the experimental curriculum could ideally have a positive impact on learning, but whether it can survive or thrive in the conditions of schooling that are so variable across sites. It is essential to know what the treatment was, whether it occurred, and if so, to what degree of intensity, fidelity, duration, and quality. In our model in Chapter 3, these factors were referred to as "implementation components." Measuring implementation can be costly for large-scale comparative studies; however, many researchers have shown that variation in implementation is a key factor in determining effectiveness. In coding the comparative studies, we identified three types of components that help to document the character of the treatment: implementation fidelity, professional development treatments, and attention to teacher effects.

Implementation Fidelity

Implementation fidelity is a measure of the basic extent of use of the curricular materials. It does not address issues of instructional quality. In some studies, implementation fidelity is synonymous with "opportunity to learn." In examining implementation fidelity, a variety of data were reported, including, most frequently, the extent of coverage of the curricular material, the consistency of the instructional approach to content in relation to the program's theory, reports of pedagogical techniques, and the length of use of the curricula at the sample sites. Other less frequently used approaches documented the calendar of curricular coverage, requested teacher feedback by textbook chapter, conducted student surveys, and gauged homework policies, use of technology, and other particular program elements. Interviews with teachers and students, classroom surveys, and observations were the most frequently used data-gathering techniques. Classroom observations were conducted infrequently in these studies, except in cases when comparative studies were combined with case studies, typically with small numbers of schools and classes where observations were conducted for long or frequent time periods. In our analysis, we coded only the presence or absence of one or more of these methods.

If the extent of implementation was used in interpreting the results, then we classified the study as having adjusted for implementation differences. Across all 63 *at least minimally methodologically adequate* studies, 44 percent reported some type of implementation fidelity measure, 3 percent reported and adjusted for it in interpreting their outcome measures, and 53 percent recorded no information on this issue. Differences among studies, by study type (NSF, UCSMP, and commercially generated), showed variation on this issue, with 46 percent of NSF reporting or adjusting for implementation, 75 percent of UCSMP, and only 11 percent of the other studies of commercial materials doing so. Of the commercial, non-UCSMP studies included, only one reported on implementation. Possibly, the evaluators for the NSF and UCSMP Secondary programs recognized more clearly that their programs demanded significant changes in practice that could affect their outcomes and could pose challenges to the teachers assigned to them.

A study by Abrams (1989) (EX)[3] on the use of Saxon algebra by ninth graders showed that concerns for implementation fidelity extend to all curricula, even those like Saxon whose methods may seem more likely to be consistent with common practice. Abrams wrote, "It was not the intent of this study to determine the effectiveness of the Saxon text when used as Saxon suggests, but rather to determine the effect of the text as it is being used in the classroom situations. However, one aspect of the research was to identify how the text is being taught, and how closely teachers adhere to its content and the recommended presentation" (p. 7). Her findings showed that for the 9 teachers and 300 students, treatment effects favoring the traditional group (using Dolciani's Algebra I textbook, Houghton Mifflin, 1980) were found on the algebra test, the algebra knowledge/skills subtest, and the problem-solving test for this population of teachers (fixed effect). No differences were found between the groups on an algebra understanding/applications subtest, overall attitude toward mathematics, mathematical self-confidence, anxiety about mathematics, or enjoyment of mathematics. She suggests that the lack of differences might be due to the ways in which teachers supplement materials, change test conditions, emphasize

---

[3]Both studies referenced in this section did not meet the criteria for inclusion in the comparative studies, but shed direct light on comparative issues of implementation. The Abrams study was omitted because it examined a program at a grade level outside the specified grade band for that curriculum. Briars and Resnick (2000) did not provide explicit comparison scores to permit one to evaluate the level of student attainment.

and deemphasize topics, use their own tests, vary the proportion of time spent on development and practice, use calculators and group work, and basically adapt the materials to their own interpretation and method. Many of these practices conflict directly with the recommendations of the authors of the materials.

A study by Briars and Resnick (2000) (EX) in Pittsburgh schools directly confronted issues relevant to professional development and implementation. Evaluators contrasted the performance of students of teachers with high and low implementation quality, and showed the results on two contrasting outcome measures, Iowa Test of Basic Skills (ITBS) and Balanced Assessment. Strong implementers were defined as those who used all of the EM components and provided student-centered instruction by giving students opportunities to explore mathematical ideas, solve problems, and explain their reasoning. Weak implementers were either not using EM or using it so little that the overall instruction in the classrooms was "hardly distinguishable from traditional mathematics instruction" (p. 8). Assignment was based on observations of student behavior in classes, the presence or absence of manipulatives, teacher questionnaires about the programs, and students' knowledge of classroom routines associated with the program.

From the identification of strong- and weak-implementing teachers, strong- and weak-implementation schools were identified as those with strong- or weak-implementing teachers in 3rd and 4th grades over two consecutive years. The performance of students with 2 years of EM experience in these settings composed the comparative samples. Three pairs of strong- and weak-implementation schools with similar demographics in terms of free and reduced-price lunch (range 76 to 93 percent), student living with only one parent (range 57 to 82 percent), mobility (range 8 to 16 percent), and ethnicity (range 43 to 98 percent African American) were identified. These students' 1st-grade ITBS scores indicated similarity in prior performance levels. Finally, evaluators predicted that if the effects were due to the curricular implementation and accompanying professional development, the effects on scores should be seen in 1998, after full implementation. Figure 5-4 shows that on the 1998 New Standards exams, placement in strong- and weak-implementation schools strongly affected students' scores. Over three years, performance in the district on skills, concepts, and problem solving rose, confirming the evaluator's predictions.

An article by McCaffrey et al. (2001) examining the interactions among instructional practices, curriculum, and student achievement illustrates the point that distinctions are often inadequately linked to measurement tools in their treatment of the terms traditional and reform teaching. In this study, researchers conducted an exploratory factor analysis that led them to create two scales for instructional practice: Reform Practices and Tradi-

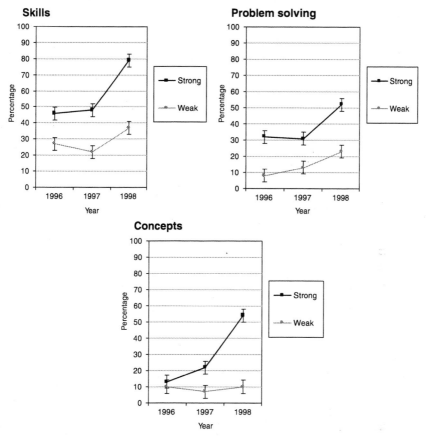

FIGURE 5-4 Percentage of students who met or exceeded the standard. District-wide grade 4 New Standards Mathematics Reference Examination (NSMRE) performance for 1996, 1997, and 1998 by level of Everyday Mathematics implementation. Percentage of students who achieved the standard. Error bars denote the 99 percent confidence interval for each data point.
SOURCE: Re-created from Briars and Resnick (2000, pp. 19-20).

tional Practices. The reform scale measured the frequency, by means of teacher report, of teacher and student behaviors associated with reform instruction and assessment practices, such as using small-group work, explaining reasoning, representing and using data, writing reflections, or performing tasks in groups. The traditional scale focused on explanations to whole classes, the use of worksheets, practice, and short-answer assessments. There was a –0.32 correlation between scores for integrated curriculum teachers. There was a 0.27 correlation between scores for traditional curriculum teachers. This shows that it is overly simplistic to think that reform and traditional practices are oppositional. The relationship among a variety of instructional practices is rather more complex as they interact with curriculum and various student populations.

Professional Development

Professional development and teacher effects were separated in our analysis from implementation fidelity. We recognized that professional development could be viewed by the readers of this report in two ways. As indicated in our model, professional development can be considered a program element or component or it can be viewed as part of the implementation process. When viewed as a program element, professional development resources are considered mandatory along with program materials. In relation to evaluation, proponents of considering professional development as a mandatory program element argue that curricular innovations, which involve the introduction of new topics, new types of assessment, or new ways of teaching, must make provision for adequate training, just as with the introduction of any new technology.

For others, the inclusion of professional development in the program elements without a concomitant inclusion of equal amounts of professional development relevant to a comparative treatment interjects a priori disproportionate treatments and biases the results. We hoped for an array of evaluation studies that might shed some empirical light on this dispute, and hence separated professional development from treatment fidelity, coding whether or not studies reported on the amount of professional development provided for the treatment and/or comparison groups. A study was coded as positive if it either reported on the professional development provided on the experimental group or reported the data on both treatments. Across all 63 *at least minimally methodologically adequate* studies, 27 percent reported some type of professional development measure, 1.5 percent reported and adjusted for it in interpreting their outcome measures, and 71.5 percent recorded no information on the issue.

A study by Collins (2002) (EX)[4] illustrates the critical and controversial role of professional development in evaluation. Collins studied the use of Connected Math over three years, in three middle schools in threat of being classified as low performing in the Massachusetts accountability system. A comparison was made between one school (School A) that engaged

---

[4]The Collins study lacked a comparison group and is coded as EX. However, it is reported as a case study.

substantively in professional development opportunities accompanying the program and two that did not (Schools B and C). In the CMP school reports (School A) totals between 100 and 136 hours of professional development were recorded for all seven teachers in grades 6 through 8. In School B, 66 hours were reported for two teachers and in School C, 150 hours were reported for eight teachers over three years. Results showed significant differences in the subsequent performance by students at the school with higher participation in professional development (School A) and it became a districtwide top performer; the other two schools remained at risk for low performance. No controls for teacher effects were possible, but the results do suggest the centrality of professional development for successful implementation or possibly suggest that the results were due to professional development rather than curriculum materials. The fact that these two interpretations cannot be separated is a problem when professional development is given to one and not the other. The effect could be due to textbook or professional development or an interaction between the two. Research designs should be adjusted to consider these issues when different conditions of professional development are provided.

Teacher Effects

These studies make it obvious that there are potential confounding factors of teacher effects. Many evaluation studies devoted inadequate attention to the variable of teacher quality. A few studies (Goodrow, 1998; Riordan and Noyce, 2001; Thompson et al., 2001; and Thompson et al., 2003) reported on teacher characteristics such as certification, length of service, experience with curricula, or degrees completed. Those studies that matched classrooms and reported by matched results rather than aggregated results sought ways to acknowledge the large variations among teacher performance and its impact on student outcomes. We coded any effort to report on possible teacher effects as one indicator of quality. Across all 63 *at least minimally methodologically adequate* studies, 16 percent reported some type of teacher effect measure, 3 percent reported and adjusted for it in interpreting their outcome measures, and 81 percent recorded no information on this issue.

One can see that the potential confounding factors of teacher effects, in terms of the provision of professional development or the measure of teacher effects, are not adequately considered in most evaluation designs. Some studies mention and give a subjective judgment as to the nature of the problem, but this is descriptive at the most. Hardly any of the studies actually do anything analytical, and because these are such important potential confounding variables, this presents a serious challenge to the efficacy of these studies. Figure 5-5 shows how attention to these factors varies

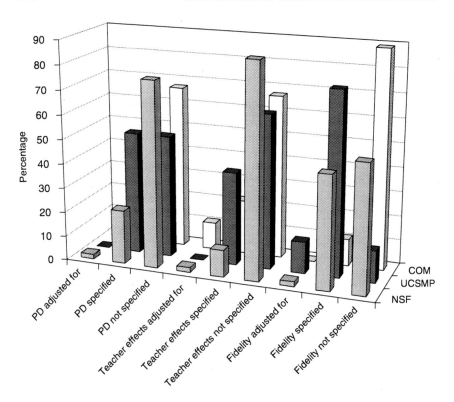

FIGURE 5-5 Treatment of implementation components by program type.
NOTE: PD = professional development.

across program categories among NSF-supported, UCSMP, and studies of commercial materials. In general, evaluations of NSF-supported studies were the most likely to measure these variables; UCSMP had the most standardized use of methods to do so across studies; and commercial material evaluators seldom reported on issues of implementation fidelity.

Identification of a Set of Outcome Measures and Forms of Disaggregation

Using the selected student outcomes identified in the program theory, one must conduct an impact assessment that refers to the design and measurement of student outcomes. In addition to selecting what outcomes should be measured within one's program theory, one must determine how these outcomes are measured, when those measures are collected, and what purpose they serve from the perspective of the participants. In the case of curricular evaluation, there are significant issues involved in how these measures are reported. To provide insight into the level of curricular validity, many evaluators prefer to report results by topic, content strand, or item cluster. These reports often present the level of specificity of outcome needed to inform curriculum designers, especially when efforts are made to document patterns of errors, distribution of results across multiple choices, or analyses of student methods. In these cases, whole test scores may mask essential differences in impact among curricula at the level of content topics, reporting only average performance.

On the other hand, many large-scale assessments depend on methods of test equating that rely on whole test scores and make comparative interpretations of different test administrations by content strands of questionable reliability. Furthermore, there are questions such as whether to present only gain scores effect sizes, how to link pretests and posttests, and how to determine the relative curricular sensitivity of various outcome measures.

The findings of comparative studies are reported in terms of the outcome measure(s) collected. To describe the nature of the database with regard to outcome measures and to facilitate our analyses of the studies, we classified each of the included studies on four outcome measure dimensions:

1. Total score reported;
2. Disaggregation of content strands, subtest, performance level, SES, or gender;
3. Outcome measure that was specific to curriculum; and
4. Use of multiple outcome measures.

Most studies reported a total score, but we did find studies that reported only subtest scores or only scores on an item-by-item basis. For example, in the Ben-Chaim et al. (1998) evaluation study of Connected Math, the authors were interested in students' proportional reasoning proficiency as a result of use of this curriculum. They asked students from eight seventh-grade classes of CMP and six seventh-grade classes from the control group to solve a variety of tasks categorized as rate and density problems. The authors provide precise descriptions of the cognitive challenges in the items; however, they do not explain if the problems written up were representative of performance on a larger set of items. A special rating form was developed to code responses in three major categories (correct answer, incorrect answer, and no response), with subcategories indicating the quality of the work that accompanied the response. No reports on reliability of coding were given. Performance on standardized tests indicated that control students' scores were slightly higher than CMP at the beginning of the year and lower at the end. Twenty-five percent of the experimental group members were interviewed about their approaches to the problems. The CMP students outperformed the control students (53 percent versus 28 percent) overall in providing the correct answers and support work, and 27 percent of the control group gave an incorrect answer or showed incorrect thinking compared to 13 percent of the CMP group. An item-level analysis permitted the researchers to evaluate the actual strategies used by the students. They reported, for example, that 82 percent of CMP students used a "strategy focused on package price, unit price, or a combination of the two; those effective strategies were used by only 56 of 91 control students (62 percent)" (p. 264).

The use of item or content strand-level comparative reports had the advantage that they permitted the evaluators to assess student learning strategies specific to a curriculum's program theory. For example, at times, evaluators wanted to gauge the effectiveness of using problems different from those on typical standardized tests. In this case, problems were drawn from familiar circumstances but carefully designed to create significant cognitive challenges, and assess how well the informal strategies approach in CMP works in comparison to traditional instruction. The disadvantages of such an approach include the use of only a small number of items and the concerns for reliability in scoring. These studies seem to represent a method of creating hybrid research models that build on the detailed analyses possible using case studies, but still reporting on samples that provide comparative data. It possibly reflects the concerns of some mathematicians and mathematics educators that the effectiveness of materials needs to be evaluated relative to very specific, research-based issues on learning and that these are often inadequately measured by multiple-choice tests. However, a decision not to report total scores led to a trade-off in the reliability and representativeness of the reported data, which must be addressed to increase the objectivity of the reports.

Second, we coded whether outcome data were disaggregated in some way. Disaggregation involved reporting data on dimensions such as content strand, subtest, test item, ethnic group, performance level, SES, and gender. We found disaggregated results particularly helpful in understanding the findings of studies that found main effects, and also in examining patterns across studies. We report the results of the studies' disaggregation by content strand in our reports of effects. We report the results of the studies' disaggregation by subgroup in our discussions of generalizability.

Third, we coded whether a study used an outcome measure that the evaluator reported as being sensitive to a particular treatment—this is a subcategory of what was defined in our framework as "curricular validity of measures." In such studies, the rationale was that readily available measures such as state-mandated tests, norm-referenced standardized tests, and college entrance examinations do not measure some of the aims of the program under study. A frequently cited instance of this was that "off the shelf" instruments do not measure well students' ability to apply their mathematical knowledge to problems embedded in complex settings. Thus, some studies constructed a collection of tasks that assessed this ability and collected data on it (Ben-Chaim et al., 1998; Huntley et al., 2000).

Finally, we recorded whether a study used multiple outcome measures. Some studies used a variety of achievement measures and other studies reported on achievement accompanied by measures such as subsequent course taking or various types of affective measures. For example, Carroll (2001, p. 47) reported results on a norm-referenced standardized achievement test as well as a collection of tasks developed in other studies.

A study by Huntley et al. (2000) illustrates how a variety of these techniques were combined in their outcome measures. They developed three assessments. The first emphasized contextualized problem solving based on items from the American Mathematical Association of Two-Year Colleges and others; the second assessment was on context-free symbolic manipulation and a third part requiring collaborative problem solving. To link these measures to the overall evaluation, they articulated an explicit model of cognition based on how one links an applied situation to mathematical activity through processes of formulation and interpretation. Their assessment strategy permitted them to investigate algebraic reasoning as an ability to use algebraic ideas and techniques to (1) mathematize quantitative problem situations, (2) use algebraic principles and procedures to solve equations, and (3) interpret results of reasoning and calculations.

In presenting their data comparing performance on Core-Plus and traditional curriculum, they presented both main effects and comparisons on subscales. Their design of outcome measures permitted them to examine differences in performance with and without context and to conclude with statements such as "This result illustrates that CPMP students perform better than control students when setting up models and solving algebraic problems presented in meaningful contexts while having access to calculators, but CPMP students do not perform as well on formal symbol-manipulation tasks without access to context cues or calculators" (p. 349). The authors go on to present data on the relationship between knowing how to plan or interpret solutions and knowing how to carry them out. The correlations between these variables were weak but significantly different (0.26 for control groups and 0.35 for Core-Plus). The advantage of using multiple measures carefully tied to program theory is that they can permit one to test fine content distinctions that are likely to be the level of adjustments necessary to fine tune and improve curricular programs.

Another interesting approach to the use of outcome measures is found in the UCSMP studies. In many of these studies, evaluators collected infor-

TABLE 5-2 Mean Percentage Correct on the Subject Tests

Treatment Group	Geometry— Standard	Geometry— UCSMP	Advanced Algebra— UCSMP
UCSMP	43.1, 44.7, 50.5[a]	51.2, 54.5[b]	56.1, 58.8, 56.1
Comparison	42.7, 45.5, 51.5	36.6, 40.8[b]	42.0, 50.1, 50.0

[a]"43.1, 44.7, 50.5" means students were correct on 43.1 percent of the total items, 44.7 percent of the fair items for UCSMP, and 50.5 percent of the items that were taught in both treatments.

[b]Too few items to report data.

SOURCES: Adapted from Thompson et al. (2001); Thompson et al. (2003).

mation from teachers' reports and chapter reviews as to whether topics for items on the posttests were taught, calling this an "opportunity to learn" measure. The authors reported results from three types of analyses: (1) total test scores, (2) fair test scores (scores reported by program but only on items on topics taught), and (3) conservative test scores (scores on common items taught in both). Table 5-2 reports on the variations across the multiple- choice test scores for the Geometry study (Thompson et al., 2003) on a standardized test, *High School Subject Tests-Geometry Form B*, and the UCSMP-constructed Geometry test, and for the Advanced Algebra Study on the UCSMP-constructed Advanced Algebra test (Thompson et al., 2001). The table shows the mean scores for UCSMP classes and comparison classes. In each cell, mean percentage correct is reported first by whole test, then by fair test, and then by conservative test.

The authors explicitly compare the items from the standard Geometry test with the items from the UCSMP test and indicate overlap and difference. They constructed their own test because, in their view, the standard test was not adequately balanced among skills, properties, and real-world uses. The UCSMP test included items on transformations, representations, and applications that were lacking in the national test. Only five items were taught by all teachers; hence in the case of the UCSMP geometry test, there is no report on a conservative test. In the Advanced Algebra evaluation, only a UCSMP-constructed test was viewed as appropriate to cover the treatment of the prior material and alignment to the goals of the new course. These data sets demonstrate the challenge of selecting appropriate outcome measures, the sensitivity of the results to those decisions, and the importance of full disclosure of decision-making processes in order to permit readers to assess the implications of the choices. The methodology utilized sought to ensure that the material in the course was covered adequately by treatment teachers while finding ways to make comparisons that reflected content coverage.

Only one study reported on its outcomes using embedded assessment items employed over the course of the year. In a study of Saxon and UCSMP, Peters (1992) (EX) studied the use of these materials with two classrooms taught by the same teacher. In this small study, he randomly assigned students to treatment groups and then measured their performance on four unit tests composed of items common to both curricula and their progress on the Orleans-Hanna Algebraic Prognosis Test.

Peters' study showed no significant difference in placement scores between Saxon and UCSMP on the posttest, but did show differences on the embedded assessment. Figure 5-6 (Peters, 1992, p. 75) shows an interesting display of the differences on a "continuum" that shows both the direction and magnitude of the differences and provides a level of concept specificity missing in many reports. This figure and a display (Figure 5-7) in a study by Senk (1991, p. 18) of students' mean scores on Curriculum A versus Curriculum B with a 10 percent range of differences marked represent two excellent means to communicate the kinds of detailed content outcome information that promises to be informative to curriculum writers, publishers, and school decision makers. In Figure 5-7, 16 items listed by number were taken from the Second International Mathematics Study. The Functions, Statistics, and Trigonometry sample averaged 41 percent correct on these items whereas the U.S. precalculus sample averaged 38 percent. As shown in the figure, differences of 10 percent or less fall inside the banded area and greater than 10 percent fall outside, producing a display that makes it easy for readers and designers to identify the relative curricular strengths and weaknesses of topics.

While we value detailed outcome measure information, we also recognize the importance of examining curricular impact on students' standardized test performance. Many developers, but not all, are explicit in rejecting standardized tests as adequate measures of the outcomes of their programs, claiming that these tests focus on skills and manipulations, that they are overly reliant on multiple-choice questions, and that they are often poorly aligned to new content emphases such as probability and statistics, transformations, use of contextual problems and functions, and process skills, such as problem solving, representation, or use of calculators. However, national and state tests are being revised to include more content on these topics and to draw on more advanced reasoning. Furthermore, these high-stakes tests are of major importance in school systems, determining graduation, passing standards, school ratings, and so forth. For this reason, if a curricular program demonstrated positive impact on such measures, we referred to that in Chapter 3 as establishing "curricular alignment with systemic factors." Adequate performance on these measures is of paramount importance to the survival of reform (to large groups of parents and

1	Terminology	−0.4
2	Addition	10.8
3	Subtraction	−5.5
4	Multiplication	3.0
5	Division	5.6
6	Linear sentences	2.5
7	Lines and distances	−10.8
8	Lines and slopes	9.6
9	Exponents and powers	1.7
10	Polynomials	17.6
11	Systems	9.5
12	Parabolas and quadratic equations	−3.0

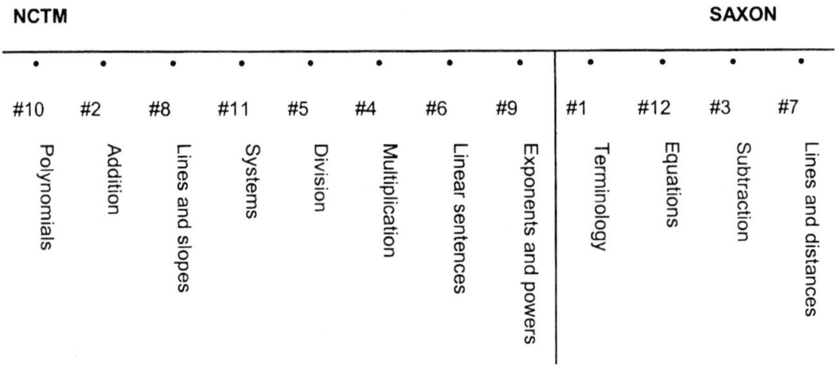

NCTM   SAXON

| #10 | #2 | #8 | #11 | #5 | #4 | #6 | #9 | #1 | #12 | #3 | #7 |
| Polynomials | Addition | Lines and slopes | Systems | Division | Multiplication | Linear sentences | Exponents and powers | Terminology | Equations | Subtraction | Lines and distances |

FIGURE 5-6 Continuum of criterion score averages for studied programs.
SOURCE: Peters (1992, p. 75).

school administrators). These examples demonstrate how careful attention to outcomes measures is an essential element of valid evaluation.

In Table 5-3, we document the number of studies using a variety of types of outcome measures that we used to code the data, and also report on the types of tests used across the studies.

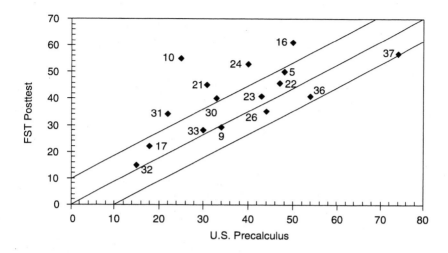

FIGURE 5-7 Achievement (percentage correct) on Second International Mathematics Study (SIMS) items by U.S. precalculus students and functions, statistics, and trigonometry (FST) students.
SOURCE: Recreated from Senk (1991, p. 18).

TABLE 5-3 Number of Studies Using a Variety of Outcome Measures by Program Type

	Total Test		Content Strands		Test Match to Program		Multiple Test	
	Yes	No	Yes	No	Yes	No	Yes	No
NSF	43	3	28	18	26	20	21	25
Commercial	8	1	4	5	2	7	2	7
UCSMP	7	1	7	1	7	1	7	1

A Choice of Statistical Tests, Including
Statistical Significance and Effect Size

In our first review of the studies, we coded what methods of statistical evaluation were used by different evaluators. Most common were t-tests; less frequently one found Analysis of Variance (ANOVA), Analysis of Co-

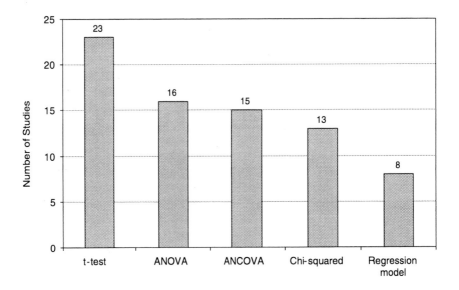

FIGURE 5-8 Statistical tests most frequently used.

variance (ANCOVA), and chi-square tests. In a few cases, results were reported using multiple regression or hierarchical linear modeling. Some used multiple tests; hence the total exceeds 63 (Figure 5-8).

One of the difficult aspects of doing curriculum evaluations concerns using the appropriate unit both in terms of the unit to be randomly assigned in an experimental study and the unit to be used in statistical analysis in either an experimental or quasi-experimental study.

For our purposes, we made the decision that unless the study concerned an intact student population such as the freshman at a single university, where a student comparison was the correct unit, we believed that for statistical tests, the unit should be at least at the classroom level. Judgments were made for each study as to whether the appropriate unit was utilized. This question is an important one because statistical significance is related to sample size, and as a result, studies that inappropriately use the student as the unit of analysis could be concluding significant differences where they are not present. For example, if achievement differences between two curricula are tested in 16 classrooms with 400 students, it will always be easier to show significant differences using scores from those 400 students than using 16 classroom means.

Fifty-seven studies used students as the unit of analysis in at least one test of significance. Three of these were coded as correct because they involved whole populations. In all, 10 studies were coded as using the

TABLE 5-4 Performance on Applied Algebra Problems with Use of Calculators, Part 1

Treatment	n	M (0-100)	SD
Control	273	34.1	14.8
CPMP	320	42.6	21.3

NOTE: t_{570}= -5.69, p < .001. All sites combined.
SOURCE: Huntley et al. (2000). Reprinted with permission.

TABLE 5-5 Reanalysis of Algebra Performance Data

Site	Site Mean		Independent Samples Difference	Dependent Sample Difference
	Control	CPMP		
1	31.7	35.5		3.8
2	26.0	49.4		23.4
3	36.7	25.2		-11.5
4	41.9	47.7		5.8
5	29.4	38.3		8.9
6	30.5	45.6		15.1
Average	32.7	40.3	7.58	7.58
Standard deviation	5.70	9.17	7.64	11.75
Standard error			4.41	4.80
		t	1.7	1.6
		p	0.116	0.175

SOURCE: Huntley et al. (2000).

correct unit of analysis; hence, 7 studies used teachers or classes, or schools. For some studies where multiple tests were conducted, a judgment was made as to whether the primary conclusions drawn treated the unit of analysis adequately. For example, Huntley et al. (2000) compared the performance of CPMP students with students in a traditional course on a measure of ability to formulate and use algebraic models to answer various questions about relationships among variables. The analysis used students as the unit of analysis and showed a significant difference, as shown in Table 5-4.

To examine the robustness of this result, we reanalyzed the data using an independent sample t-test and a matched pairs t-test with class means as the unit of analysis in both tests (Table 5-5). As can be seen from the analyses, in neither statistical test was the difference between groups found to be significantly different (p < .05), thus emphasizing the importance of using the correct unit in analyzing the data.

Reanalysis of student-level data using class means will not always result

TABLE 5-6 Mean Percentage Correct on Entire Multiple-Choice Posttest: Second Edition and Non-UCSMP

School	Pair	UCSMP Second Edition			
Code	ID	n	Mean	SD	OTL
J	18	18	60.8	9.0	100
J	19	11	58.8	13.5	100
K	20	22	63.8	13.0	94
K	21	16	64.8	14.0	94
L	22	19	57.6	16.9	92
L	23	13	44.7	11.2	92
M	24	29	58.4	12.7	92
M	25	22	39.6	13.5	92
Overall		150	56.1	15.4	

NOTE: The mean is the mean percentage correct on a 36-item multiple-choice posttest. The OTL is the percentage of the items for which teachers reported their students had the opportunity to learn the needed content. Underline indicates statistically significant differences between the mean percentage correct for each pair.

in a change in finding. Furthermore, using class means as the unit of analysis does not suggest that significant differences will not be found. For example, a study by Thompson et al. (2001) compared the performance of UCSMP students with the performance of students in a more traditional program across several measures of achievement. They found significant differences between UCSMP students and the non-UCSMP students on several measures. Table 5-6 shows results of an analysis of a multiple-choice algebraic posttest using class means as the unit of analysis. Significant differences were found in five of eight separate classroom comparisons, as shown in the table. They also found a significant difference using a matched-pairs t-test on class means.

The lesson to be learned from these reanalyses is that the choice of unit of analysis and the way the data are aggregated can impact study findings in important ways including the extent to which these findings can be generalized. Thus it is imperative that evaluators pay close attention to such considerations as the unit of analysis and the way data are aggregated in the design, implementation, and analysis of their studies.

Non-UCSMP[a]							
n	Mean	SD	OTL	SE	t	df	p
14	55.2	10.2	69	3.40	1.65	30	0.110
15	53.7	11.0	69	4.81	1.06	24	0.299
24	45.9	10.0	72	3.41	5.22	44	0.000
23	43.0	11.9	72	4.16	5.23	37	0.000
20	38.8	9.1	75	4.32	4.36	37	0.000
15	38.3	11.0	75	4.20	1.52	26	0.140
22	37.8	13.8	47	3.72	5.56	49	0.000
23	30.8	9.9	47	3.52	2.51	43	0.016
156	42.0	13.1					

[a]A matched-pairs t-test indicates that the differences between the two curricula are significant.
($\bar{X} = 13.125, S_{\bar{X}} = 7.281, t = 5.099, p = 0.0014$)
SOURCE: Thompson et al. (2001). Reprinted with permission.

Second, effect size has become a relatively common and standard way of gauging the practical significance of the findings. Statistical significance only indicates whether the main-level differences between two curricula are large enough to not be due to chance, assuming they come from the same population. When statistical differences are found, the question remains as to whether such differences are large enough to consider. Because any innovation has its costs, the question becomes one of cost-effectiveness: Are the differences in student achievement large enough to warrant the costs of change? Quantifying the practical effect once statistical significance is established is one way to address this issue. There is a statistical literature for doing this, and for the purposes of this review, the committee simply noted whether these studies have estimated such an effect. However, the committee further noted that in conducting meta-analyses across these studies, effect size was likely to be of little value. These studies used an enormous variety of outcome measures, and even using effect size as a means to standardize units across studies is not sensible when the measures in each study address such a variety of topics, forms of reasoning, content levels, and assessment strategies.

We note very few studies drew upon the advances in methodologies employed in modeling, which include causal modeling, hierarchical linear modeling (Bryk and Raudenbush, 1992; Bryk et al., 1993), and selection bias modeling (Heckman and Hotz, 1989). Although developing detailed specifications for these approaches is beyond the scope of this review, we wish to emphasize that these methodological advances should be considered within future evaluation designs.

Results and Limitations to Generalizability Resulting from Design Constraints

One also must consider what generalizations can be drawn from the results (Campbell and Stanley, 1966; Caporaso and Roos, 1973; and Boruch, 1997). Generalization is a matter of external validity in that it determines to what populations the study results are likely to apply. In designing an evaluation study, one must carefully consider, in the selection of units of analysis, how various characteristics of those units will affect the generalizability of the study. It is common for evaluators to conflate issues of representativeness for the purpose of generalizability (external validity) and comparativeness (the selection of or adjustment for comparative groups [internal validity]). Not all studies must be representative of the population served by mathematics curricula to be internally valid. But, to be generalizable beyond restricted communities, representativeness must be obtained by the random selection of the basic units. Clearly specifying such limitations to generalizability is critical. Furthermore, on the basis of equity considerations, one must be sure that if overall effectiveness is claimed, that the studies have been conducted and analyzed with reference of all relevant subgroups.

Thus, depending on the design of a study, its results may be limited in generalizability to other populations and circumstances. We identified four typical kinds of limitations on the generalizability of studies and coded them to determine, on the whole, how generalizable the results across studies might be.

First, there were studies whose designs were limited by the ability or performance level of the students in the samples. It was not unusual to find that when new curricula were implemented at the secondary level, schools kept in place systems of tracking that assigned the top students to traditional college-bound curriculum sequences. As a result, studies either used comparative groups who were matched demographically but less skilled than the population as a whole, in relation to prior learning, or their results compared samples of less well-prepared students to samples of students with stronger preparations. Alternatively, some studies reported on the effects of curricula reform on gifted and talented students or on college-attending students. In these cases, the study results would also limit the generalizability of the results to similar populations. Reports using limited samples of students' ability and prior performance levels were coded as a limitation to the generalizability of the study.

For example, Wasman (2000) conducted a study of one school (six teachers) and examined the students' development of algebraic reasoning after one (n=100) and two years (n=73) in CMP. In this school, the top 25 percent of the students are counseled to take a more traditional algebra course, so her experimental sample, which was 61 percent white, 35 percent African American, 3 percent Asian, and 1 percent Hispanic, consisted of the lower 75 percent of the students. She reported on the student performance on the Iowa Algebraic Aptitude Test (IAAT) (1992), in the subcategories of interpreting information, translating symbols, finding relationships, and using symbols. Results for Forms 1 and 2 of the test, for the experimental and norm group, are shown in Table 5-7 for 8th graders.

In our coding of outcomes, this study was coded as showing no significant differences, although arguably its results demonstrate a positive set of

TABLE 5-7 Comparing Iowa Algebraic Aptitude Test (IAAT) Mean Scores of the Connected Mathematics Project Forms 1 and 2 to the Normative Group (8th Graders)

	Interpreting Information	Translating Symbols	Finding Relationships	Using Symbols	Total
CMP: Form 1 7th (n=51)	9.35 (3.36)	8.22 (3.44)	9.90 (3.26)	8.65 (3.12)	36.12 (11.28)
CMP: Form 1 8th (n=41)	9.76 (3.89)	8.56 (3.64)	9.41 (4.13)	8.27 (3.74)	36.00 (13.65)
Norm: Form 1 (n=2,467)	10.03 (3.35)	9.55 (2.89)	9.14 (3.59)	8.87 (3.19)	37.59 (10.57)
CMP: Form 2 7th (n=49)	9.41 (4.05)	7.82 (3.03)	9.29 (3.57)	7.65 (3.35)	34.16 (11.47)
CMP: Form 2 8th (n=32)	11.28 (3.74)	8.66 (3.81)	10.94 (3.79)	9.81 (3.64)	40.69 (12.94)
Norm: Form 2 (n=2,467)	10.63 (3.78)	8.58 (2.91)	8.67 (3.84)	9.19 (3.17)	37.07 (11.05)

NOTE: Parentheses indicate standard deviation.
SOURCE: Adapted from Wasman (2000).

outcomes as the treatment group was weaker than the control group. Had the researcher used a prior achievement measure and a different statistical technique, significance might have been demonstrated, although potential teacher effects confound interpretations of results.

A second limitation to generalizability was when comparative studies resided entirely at curriculum pilot site locations, where such sites were developed as a means to conduct formative evaluations of the materials with close contact and advice from teachers. Typically, pilot sites have unusual levels of teacher support, whether it is in the form of daily technical support in the use of materials or technology or increased quantities of professional development. These sites are often selected for study because they have established cooperative agreements with the program developers and other sources of data, such as classroom observations, are already available. We coded whether the study was conducted at a pilot site to signal potential limitations in generalizability of the findings.

Third, studies were also coded as being of limited generalizability if they failed to disaggregate their data by socioeconomic class, race, gender, or some other potentially significant sources of restriction on the claims. We recorded the categories in which disaggregation occurred and compiled their frequency across the studies. Because of the need to open the pipeline to advanced study in mathematics by members of underrepresented groups, we were particularly concerned about gauging the extent to which evaluators factored such variables into their analysis of results and not just in terms of the selection of the sample.

Of the 46 included studies of NSF-supported curricula, 19 disaggregated their data by student subgroup. Nine of 17 studies of commercial materials disaggregated their data. Figure 5-9 shows the number of studies that disaggregated outcomes by race or ethnicity, SES, gender, LEP, special education status, or prior achievement. Studies using multiple categories of disaggregation were counted multiple times by program category.

The last category of restricted generalization occurred in studies of limited sample size. Although such studies may have provided more in-depth observations of implementation and reports on professional development factors, the smaller numbers of classrooms and students in the study would limit the extent of generalization that could be drawn from it. Figure 5-10 shows the distribution of sizes of the samples in terms of numbers of students by study type.

Summary of Results by Student Achievement Among Program Types

We present the results of the studies as a means to further investigate their methodological implications. To this end, for each study, we counted across outcome measures the number of findings that were positive, nega-

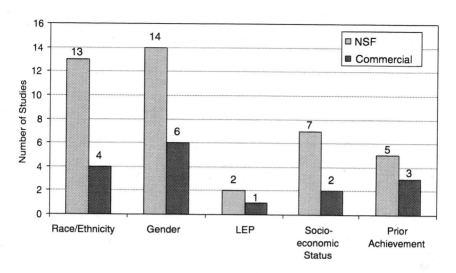

FIGURE 5-9 Disaggregation of subpopulations.

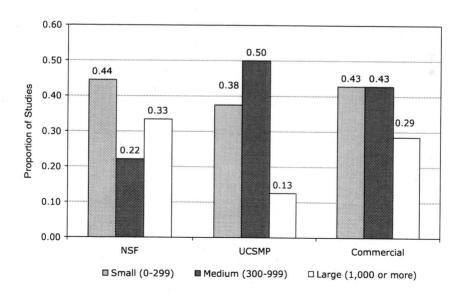

FIGURE 5-10 Proportion of studies by sample size and program.

tive, or indeterminate (no significant difference) and then calculated the proportion of each. We represented the calculation of each study as a triplet (a, b, c) where *a* indicates the proportion of the results that were positive and statistically significantly stronger than the comparison program, *b* indicates the proportion that were negative and statistically significantly weaker than the comparison program, and *c* indicates the proportion that showed no significant difference between the treatment and the comparative group. For studies with a single outcome measure, without disaggregation by content strand, the triplet is always composed of two zeros and a single one. For studies with multiple measures or disaggregation by content strand, the triplet is typically a set of three decimal values that sum to one. For example, a study with one outcome measure in favor of the experimental treatment would be coded (1, 0, 0), while one with multiple measures and mixed results more strongly in favor of the comparative curriculum might be listed as (.20, .50, .30). This triplet would mean that for 20 percent of the comparisons examined, the evaluators reported statistically significant positive results, for 50 percent of the comparisons the results were statistically significant in favor of the comparison group, and for 30 percent of the comparisons no significant difference were found. Overall, the mean score on these distributions was (.54, .07, .40), indicating that across all the studies, 54 percent of the comparisons favored the treatment, 7 percent favored the comparison group, and 40 percent showed no significant difference. Table 5-8 shows the comparison by curricular program types. We present the results by individual program types, because each program type relies on a similar program theory and hence could lead to patterns of results that would be lost in combining the data. If the studies of commercial materials are all grouped together to include UCSMP, their pattern of results is (.38, .11, .51). Again we emphasize that due to our call for increased methodological rigor and the use of multiple methods, this result is not sufficient to establish the curricular effectiveness of these programs as a whole with adequate certainty.

We caution readers that these results are summaries of the results presented across a set of evaluations that meet only the standard of *at least*

TABLE 5-8 Comparison by Curricular Program Types

Proportion of Results That Are:	NSF-Supported n=46	UCSMP n=8	Commercially Generated n=9
In favor of treatment	.591	.491	.285
In favor of comparison	.055	.087	.130
Show no significant difference	.354	.422	.585

minimally methodologically adequate. Calculations of statistical significance of each program's results were reported by the evaluators; we have made no adjustments for weaknesses in the evaluations such as inappropriate use of units of analysis in calculating statistical significance. Evaluations that consistently used the correct unit of analysis, such as UCSMP, could have fewer reports of significant results as a consequence. Furthermore, these results are not weighted by study size. Within any study, the results pay no attention to comparative effect size or to the established credibility of an outcome measure. Similarly, these results do not take into account differences in the populations sampled, an important consideration in generalizing the results. For example, using the same set of studies as an example, UCSMP studies used volunteer samples who responded to advertisements in their newsletters, resulting in samples with disproportionately Caucasian subjects from wealthier schools compared to national samples. As a result, we would suggest that these results are useful only as baseline data for future evaluation efforts. Our purpose in calculating these results is to permit us to create filters from the critical decision points and test how the results change as one applies more rigorous standards.

Given that none of the studies adequately addressed all of the critical criteria, we do not offer these results as definitive, only suggestive—a hypothesis for further study. In effect, given the limitations of time and support, and the urgency of providing advice related to policy, we offer this filtering approach as an informal meta-analytic technique sufficient to permit us to address our primary task, namely, evaluating the quality of the evaluation studies.

This approach reflects the committee's view that to deeply understand and improve methodology, it is necessary to scrutinize the results and to determine what inferences they provide about the conduct of future evaluations. Analogous to debates on consequential validity in testing, we argue that to strengthen methodology, one must consider what current methodologies are able (or not able) to produce across an entire series of studies. The remainder of the chapter is focused on considering in detail what claims are made by these studies, and how robust those claims are when subjected to challenge by alternative hypothesis, filtering by tests of increasing rigor, and examining results and patterns across the studies.

Alternative Hypotheses on Effectiveness

In the spirit of scientific rigor, the committee sought to consider rival hypotheses that could explain the data. Given the weaknesses in the designs generally, often these alternative hypotheses cannot be dismissed. However, we believed that only after examining the configuration of results and alternative hypotheses can the next generation of evaluations be better informed and better designed. We began by generating alternative hypotheses to explain the positive directionality of the results in favor of experimental groups. Alternative hypotheses included the following:

- The teachers in the experimental groups tended to be self-selecting early adopters, and thus able to achieve effects not likely in regular populations.
- Changes in student outcomes reflect the effects of professional development instruction, or level of classroom support (in pilot sites), and thus inflate the predictions of effectiveness of curricular programs.
- Hawthorne effect (Franke and Kaul, 1978) occurs when treatments are compared to everyday practices, due to motivational factors that influence experimental participants.
- The consistent difference is due to the coherence and consistency of a single curricular program when compared to multiple programs.
- The significance level is only achieved by the use of the wrong unit of analysis to test for significance.
- Supplemental materials or new teaching techniques produce the results and not the experimental curricula.
- Significant results reflect inadequate outcome measures that focus on a restricted set of activities.
- The results are due to evaluator bias because too few evaluators are independent of the program developers.

At the same time, one could argue that the results actually underestimate the performance of these materials and are conservative measures, and their alternative hypotheses also deserve consideration:

- Many standardized tests are not sensitive to these curricular approaches, and by eliminating studies focusing on affect, we eliminated a key indicator of the appeal of these curricula to students.
- Poor implementation or increased demands on teachers' knowledge dampens the effects.
- Often in the experimental treatment, top-performing students are missing as they are advised to take traditional sequences, rendering the samples unequal.
- Materials are not well aligned with universities and colleges because tests for placement and success in early courses focus extensively on algebraic manipulation.
- Program implementation has been undercut by negative publicity and the fears of parents concerning change.

There are also a number of possible hypotheses that may be affecting the results in either direction, and we list a few of these:

- Examining the role of the teacher in curricular decision making is an important element in effective implementation, and design mandates of evaluation design make this impossible (and the positives and negatives or single- versus dual-track curriculum as in Lundin, 2001).
- Local tests that are sensitive to the curricular effects typically are not mandatory and hence may lead to unpredictable performance by students.
- Different types and extent of professional development may affect outcomes differentially.
- Persistence or attrition may affect the mean scores and are often not considered in the comparative analyses.

One could also generate reasons why the curricular programs produced results showing no significance when one program or the other is actually more effective. This could include high degrees of variability in the results, samples that used the correct unit of analysis but did not obtain consistent participation across enough cases, implementation that did not show enough fidelity to the measures, or outcome measures insensitive to the results. Again, subsequent designs should be better informed by these findings to improve the likelihood that they will produce less ambiguous results and replication of studies could also give more confidence in the findings.

It is beyond the scope of this report to consider each of these alternative hypotheses separately and to seek confirmation or refutation of them. However, in the next section, we describe a set of analyses carried out by the committee that permits us to examine and consider the impact of various critical evaluation design decisions on the patterns of outcomes across sets of studies. A number of analyses shed some light on various alternative hypotheses and may inform the conduct of future evaluations.

Filtering Studies by Critical Decision Points to Increase Rigor

In examining the comparative studies, we identified seven critical decision points that we believed would directly affect the rigor and efficacy of the study design. These decision points were used to create a set of 16 filters. These are listed as the following questions:

1. Was there a report on comparability relative to SES?
2. Was there a report on comparability of samples relative to prior knowledge?
3. Was there a report on treatment fidelity?
4. Was professional development reported on?

    5. Was the comparative curriculum specified?

    6. Was there any attempt to report on teacher effects?

    7. Was a total test score reported?

    8. Was total test score(s) disaggregated by content strand?

    9. Did the outcome measures match the curriculum?

   10. Were multiple tests used?

   11. Was the appropriate unit of analysis used in their statistical tests?

   12. Did they estimate effect size for the study?

   13. Was the generalizability of their findings limited by use of a restricted range of ability levels?

   14. Was the generalizability of their findings limited by use of pilot sites for their study?

   15. Was the generalizability of their findings limited by not disaggregating their results by subgroup?

   16. Was the generalizability of their findings limited by use of small sample size?

The studies were coded to indicate if they reported having addressed these considerations. In some cases, the decision points were coded dichotomously as present or absent in the studies, and in other cases, the decision points were coded trichotomously, as description presented, absent, or statistically adjusted for in the results. For example, a study may or may not report on the comparability of the samples in terms of race, ethnicity, or socioeconomic status. If a report on SES was given, the study was coded as "present" on this decision; if a report was missing, it was coded as "absent"; and if SES status or ethnicity was used in the analysis to actually adjust outcomes, it was coded as "adjusted for." For each coding, the table that follows reports the number of studies that met that condition, and then reports on the mean percentage of statistically significant results, and results showing no significant difference for that set of studies. A significance test is run to see if the application of the filter produces changes in the probability that are significantly different.[5]

In the cases in which studies are coded into three distinct categories—present, absent, and adjusted for—a second set of filters is applied. First, the studies coded as present or adjusted for are combined and compared to those coded as absent; this is what we refer to as a weak test of the rigor of the study. Second, the studies coded as present or absent are combined and compared to those coded as adjusted for. This is what we refer to as a strong test. For dichotomous codings, there can be as few as three compari-

---

[5]The significance test used was a chi-square not corrected for discontinuity.

sons, and for trichotomous codings, there can be nine comparisons with accompanying tests of significance. Trichotomous codes were used for adjustments for SES and prior knowledge, examining treatment fidelity, professional development, teacher effects, and reports on effect sizes. All others were dichotomous.

NSF Studies and the Filters

For example, there were 11 studies of NSF-supported curricula that simply reported on the issues of SES in creating equivalent samples for comparison, and for this subset the mean probabilities of getting positive, negative, or results showing no significant difference were (.47, .10, .43). If no report of SES was supplied (n= 21), those probabilities become (.57, .07, .37), indicating an increase in positive results and a decrease in results showing no significant difference. When an adjustment is made in outcomes based on differences in SES (n=14), the probabilities change to (.72, .00, .28), showing a higher likelihood of positive outcomes. The probabilities that result from filtering should always be compared back to the overall results of (.59, .06, .35) (see Table 5-8) so as to permit one to judge the effects of more rigorous methodological constraints. This suggests that a simple report on SES without adjustment is least likely to produce positive outcomes; that is, no report produces the outcomes next most likely to be positive and studies that adjusted for SES tend to have a higher proportion of their comparisons producing positive results.

The second method of applying the filter (the weak test for rigor) for the treatment of the adjustment of SES groups compares the probabilities when a report is either given or adjusted for compared to when no report is offered. The combined percentage of a positive outcome of a study in which SES is reported or adjusted for is (.61, .05, .34), while the percentage for no report remains as reported previously at (.57, .07, .37). A final filter compares the probabilities of the studies in which SES is adjusted for with those that either report it only or do not report it at all. Here we compare the percentage of (.72, .00, .28) to (.53, .08, .37) in what we call a strong test. In each case we compared the probability produced by the whole group to those of the filtered studies and conducted a test of the differences to determine if they were significant. These differences were not significant. These findings indicate that to date, with this set of studies, there is no statistically significant difference in results when one reports or adjusts for changes in SES. It appears that by adjusting for SES, one sees increases in the positive results, and this result deserves a closer examination for its implications should it prove to hold up over larger sets of studies.

We ran tests that report the impact of the filters on the number of studies, the percentage of studies, and the effects described as probabilities for each of the three study categories, NSF-supported and commercially generated with UCSMP included. We claim that when a pattern of probabilities of results does not change after filtering, one can have more confidence in that pattern. When the pattern of results changes, there is a need for an explanatory hypothesis, and that hypothesis can shed light on experimental design. We propose that this "filtering process" constitutes a test of the robustness of the outcome measures subjected to increasing degrees of rigor by using filtering.

Results of Filtering on Evaluations of NSF-Supported Curricula

For the NSF-supported curricular programs, out of 15 filters, 5 produced a probability that differed significantly at the p<.1 level. The five filters were for treatment fidelity, specification of control group, choosing the appropriate statistical unit, generalizability for ability, and generalizability based on disaggregation by subgroup. For each filter, there were from three to nine comparisons, as we examined how the probabilities of outcomes change as tests were more stringent and across the categories of positive results, negative results, and results with no significant differences. Out of a total of 72 possible tests, only 11 produced a probability that differed significantly at the p < .1 level. With 85 percent of the comparisons showing no significant difference after filtering, we suggest the results of the studies were relatively robust in relation to these tests. At the same time, when rigor is increased for the five filters just listed, the results become generally more ambiguous and signal the need for further research with more careful designs.

Studies of Commercial Materials and the Filters

To ensure enough studies to conduct the analysis (n=17), our filtering analysis of the commercially generated studies included UCSMP (n=8). In this case, there were six filters that produced a probability that differed significantly at the p < .1 level. These were treatment fidelity, disaggregation by content, use of multiple tests, use of effect size, generalizability by ability, and generalizability by sample size. In this case, because there were no studies in some possible categories, there were a total of 57 comparisons, and 9 displayed significant differences in the probabilities after filtering at the p < .1 level. With 84 percent of the comparisons showing no significant difference after filtering, we suggest the results of the studies were relatively robust in relation to these tests. Table 5-9 shows the cases in which significant differences were recorded.

Impact of Treatment Fidelity on Probabilities

A few of these differences are worthy of comment. In the cases of both the NSF-supported and commercially generated curricula evaluation studies, studies that reported treatment fidelity differed significantly from those that did not. In the case of the studies of NSF-supported curricula, it appeared that a report or adjustment on treatment fidelity led to proportions with less positive effects and more results showing no significant differences. We hypothesize that this is partly because larger studies often do not examine actual classroom practices, but can obtain significance more easily due to large sample sizes.

In the studies of commercial materials, the presence or absence of measures of treatment fidelity worked differently. Studies reporting on or adjusting for treatment fidelity tended to have significantly higher probabilities in favor of experimental treatment, less positive effects in fewer of the comparative treatments, and more likelihood of results with no significant differences. We hypothesize, and confirm with a separate analysis, that this is because UCSMP frequently reported on treatment fidelity in their designs while study of Saxon typically did not, and the change represents the preponderance of these different curricular treatments in the studies of commercially generated materials.

Impact of Identification of Curricular Program on Probabilities

The significant differences reported under specificity of curricular comparison also merit discussion for studies of NSF-supported curricula. When the comparison group is not specified, a higher percentage of mean scores in favor of the experimental curricula is reported. In the studies of commercial materials, a failure to name specific curricular comparisons also produced a higher percentage of positive outcomes for the treatment, but the difference was not statistically significant. This suggests the possibility that when a specified curriculum is compared to an unspecified curriculum, reports of impact may be inflated. This finding may suggest that in studies of effectiveness, specifying comparative treatments would provide more rigorous tests of experimental approaches.

When studies of commercial materials disaggregate their results of content strands or use multiple measures, their reports of positive outcomes increase, the negative outcomes decrease, and in one case, the results show no significant differences. Percentage of significant difference was only recorded in one comparison within each one of these filters.

TABLE 5-9 Cases of Significant Differences

Test	Type of Comparison	Category Code	N=	Probabilities Before Filter	p=
NSF STUDIES					
Treatment fidelity	Simple compare	Specified	21	.51, .02, .47*	*p =.049
	Not specified		24	.68, .09, .23*	
	Adjusted for		1	.25, .00, .75	
Treatment fidelity	Strong test	Adjusted for	22	.49*, .02, .49**	*p=.098
	Reported or				**p=.019
	not specified		24	.68*, .09, .23**	
Control group specified	Simple compare	Specified	8	.33*, .00, .66**	*p=.033
		Not specified	38	.65*, .07, .29**	**p=.008
Appropriate unit of analysis	Simple compare	Correct	5	.30*, .40**, .30	*p=.069
		Incorrect	41	.63*, .01**, .36	**p=.000
Generalizability by ability	Simple compare	Limited	5	.22*, .41**, .37	*p=.019
		Not limited	41	.64*, .01**, .35	**p=.000
Generalizability by disaggregated subgroup	Simple compare	Limited	28	.48*, .09, .43**	*p=.013
		Not limited	18	.76*, .00, .24**	**p=.085

COMM STUDIES

			N		
Treatment fidelity	Simple compare	Reported	7	.53, .37*, .20	*p=.032
		Not specified	9	.26, .67*, .11	
		Adjusted for	1	.45, .00*, .55	
Treatment fidelity	Weak test	Adjusted for or	8	.52, .33, .25*	*p=.087
		Reported versus	9	.26, .67, .11*	
		Not specified			
Outcomes disaggregated by content strand	Simple compare	Reported	11	.50, .37, .22*	*p=.052
		Not reported	6	.17, .77, .10*	
Outcomes using multiple tests	Simple compare	Yes	9	.55*, .35, .19	*p=.076
		No	8	.20*, .68, .20	
Effect size reported	Simple compare	Yes	3	.72, .05, .29*	*p=.029
		No	14	.31, .61, .16*	
Generalization by ability	Simple compare	Limited	4	.23, .41*, .32	*p=.004
		Not limited	14	.42, .53, .09	
Generalization by sample size	Simple compare	Limited	6	.57, .23, .27*	*p=.036
		Not limited	11	.28, .66, .10*	

NOTE: In the comparisons shown, only the comparisons marked by an asterisk showed significant differences at $p<.1$. Probabilities are estimated for each significant difference.

Impact of Units of Analysis on Probabilities[6]

For the evaluations of the NSF-supported materials, a significant difference was reported on the outcomes for the studies that used the correct unit of analysis compared to those that did not. The percentage for those with the correct unit were (.30, .40, .30) compared to (.63, .01, .36) for those that used the incorrect result. These results suggest that our prediction that using the correct unit of analysis would decrease the percentage of positive outcomes is likely to be correct. It also suggests that the most serious threat to the apparent conclusions of these studies comes from selecting an incorrect unit of analysis. It causes a decrease in favorable results, making the results more ambiguous, but never reverses the direction of the effect. This is a concern that merits major attention in the conduct of further studies.

For the commercially generated studies, most of the ones coded with the correct unit of analysis were UCSMP studies. Because of the small number of studies involved, we could not break out from the overall filtering of studies of commercial materials, but report this issue to assist readers in interpreting the relative patterns of results.

Impact of Generalizability on Probabilities

Both types of studies yielded significant differences for some of the comparisons coded as restrictions to generalizability. Investigating these is important in order to understand the effects of these curricular programs on different subpopulations of students. In the case of the studies of commercially generated materials, significantly different results occurred in the categories of ability and sample size. In the studies of NSF-supported materials, the significant differences occurred in ability and disaggregation by subgroups.

In relation to generalizability, the studies of NSF-supported curricula reported significantly more positive results in favor of the treatment when they included all students. Because studies coded as "limited by ability" were restricted either by focusing only on higher achieving students or on lower achieving students, we sorted these two groups. For higher performing students (n=3), the probabilities of effects were (.11, .67, .22). For lower

---

[6]It should be noted that of the five studies in which the correct unit of analysis was used, two of these were population studies of freshmen entering college, and these reported few results in favor of the experimental treatments. However, the high proportion of these studies involving college students may skew this particular result relative to the preponderance of other studies involving K-12 students.

performing students (n=2), the probabilities were (.39, .025, .59). The first two comparisons are significantly different at p < .05. These findings are based on only a total of five studies, but they suggest that these programs may be serving the weaker ability students more effectively than the stronger ability students, serving both less well than they serve whole heterogeneous groups. For the studies of commercial materials, there were only three studies that were restricted to limited populations. The results for those three studies were (.23, .41, .32) and for all students (n=14) were (.42, .53, .09). These studies were significantly different at p = .004. All three studies included UCSMP and one also included Saxon and was limited by serving primarily high-performing students. This means both categories of programs are showing weaker results when used with high-ability students.

Finally, the studies on NSF-supported materials were disaggregated by subgroups for 28 studies. A complete analysis of this set follows, but the studies that did not report results disaggregated by subgroup generated probabilities of results of (.48, .09, .43) whereas those that did disaggregate their results reported (.76, 0, .24). These gains in positive effects came from significant losses in reporting no significant differences. Studies of commercial materials also reported a small decrease in likelihood of negative effects for the comparison program when disaggregation by subgroup is reported offset by increases in positive results and results with no significant differences, although these comparisons were not significantly different. A further analysis of this topic follows.

Overall, these results suggest that increased rigor seems to lead in general to less strong outcomes, but never reports of completely contrary results. These results also suggest that in recommending design considerations to evaluators, there should be careful attention to having evaluators include measures of treatment fidelity, considering the impact on all students as well as one particular subgroup; using the correct unit of analysis; and using multiple tests that are also disaggregated by content strand.

Further Analyses

We conducted four further analyses: (1) an analysis of the outcome probabilities by test type; (2) content strands analysis; (3) equity analysis; and (4) an analysis of the interactions of content and equity by grade band. Careful attention to the issues of content strand, equity, and interaction is essential for the advancement of curricular evaluation. Content strand analysis provides the detail that is often lost by reporting overall scores; equity analysis can provide essential information on what subgroups are adequately served by the innovations, and analysis by content and grade level can shed light on the controversies that evolve over time.

Analysis by Test Type

Different studies used varied combinations of outcome measures. Because of the importance of outcome measures on test results, we chose to examine whether the probabilities for the studies changed significantly across different types of outcome measures (national test, local test). The most frequent use of tests across all studies was a combination of national and local tests (n=18 studies), a local test (n=16), and national tests (n=17). Other uses of test combinations were used by three studies or less. The percentages of various outcomes by test type in comparison to all studies are described in Table 5-10.

These data (Table 5-11) suggest that national tests tend to produce less positive results, and with the resulting gains falling into results showing no significant differences, suggesting that national tests demonstrate less curricular sensitivity and specificity.

TABLE 5-10 Percentage of Outcomes by Test Type

Test Type	National/Local	Local Only	National Only	All Studies
All studies	(.48, .18, .34) n=18	(.63, .03, .34) n=16	(.31, .05, .64) n= 3	(.54, .07, .40) n=63

NOTE: The first set of numbers in the parenthesis represent the percentage of outcomes that are positive, the second set of numbers in the parenthesis represent the percentage of outcomes that are negative, and the third set of numbers represent the percentage of outcomes that are nonsignificant.

TABLE 5-11 Percentage of Outcomes by Test Type and Program Type

Test Type	National/Local	Local Only	National Only	All Studies
NSF effects	(.52, .15, .34) n=14	(.57, .03, .39) n=14	(.44, .00, .56) n=4	(.59, .06, .35) n=46
UCSMP effects	(.41, .18, .41) n=3	***	***	(.49, .09, .42) n=8
Commercial effects	**	**	(.29, .08, .63) n=8	(.29, .13, .59) n=9

NOTE: The first set of numbers in the parenthesis represent the percentage of outcomes that are positive, the second set of numbers represent the percentage of outcomes that are negative, and the third set of numbers represent the percentage of outcomes that are nonsignificant.

TABLE 5-12 Number of Studies That Disaggregated by Content Strand

Program Type	Elementary	Middle	High School	Total
NSF-supported	14	6	9	29
Commercially generated	0	4	5	9

Content Strand

Curricular effectiveness is not an all-or-nothing proposition. A curriculum may be effective in some topics and less effective in others. For this reason, it is useful for evaluators to include an analysis of curricular strands and to report on the performance of students on those strands. To examine this issue, we conducted an analysis of the studies that reported their results by content strand. Thirty-eight studies did this; the breakdown is shown in Table 5-12 by type of curricular program and grade band.

To examine the evaluations of these content strands, we began by listing all of the content strands reported across studies as well as the frequency of report by the number of studies at each grade band. These results are shown in Figure 5-11, which is broken down by content strand, grade level, and program type.

Although there are numerous content strands, some of them were reported on infrequently. To allow the analysis to focus on the key results from these studies, we separated out the most frequently reported on strands, which we call the "major content strands." We defined these as strands that were examined in at least 10 percent of the studies. The major content strands are marked with an asterisk in the Figure 5-11. When we conduct analyses across curricular program types or grade levels, we use these to facilitate comparisons.

A second phase of our analysis was to examine the performance of students by content strand in the treatment group in comparison to the control groups. Our analysis was conducted across the major content strands at the level of NSF-supported versus commercially generated, initially by all studies and then by grade band. It appeared that such analysis permitted some patterns to emerge that might prove helpful to future evaluators in considering the overall effectiveness of each approach. To do this, we then coded the number of times any particular strand was measured across all studies that disaggregated by content strand. Then, we coded the proportion of times that this strand was reported as favoring the experimental treatment, favoring the comparative curricula, or showing no significant difference. These data are presented across the major content strands for the NSF-supported curricula (Figure 5-12) and the commercially generated curricula, (Figure 5-13) (except in the case of the elemen-

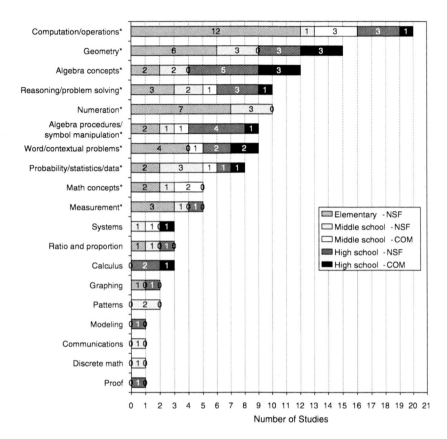

FIGURE 5-11 Study counts for all content strands.

tary curricula where no data were available) in the forms of percentages, with the frequencies listed in the bars.

The presentation of results by strands must be accompanied by the same restrictions as stated previously. These results are based on studies identified as *at least minimally methodologically adequate*. The quality of the outcome measures in measuring the content strands has not been examined. Their results are coded in relation to the comparison group in the study and are indicated as statistically in favor of the program, as in favor of the comparative program, or as showing no significant differences. The results are combined across studies with no weighting by study size. Their results should be viewed as a means for the identification of topics for potential future study. It is completely possible that a refinement of methodologies may affect the future patterns of results, so the results are to be viewed as tentative and suggestive.

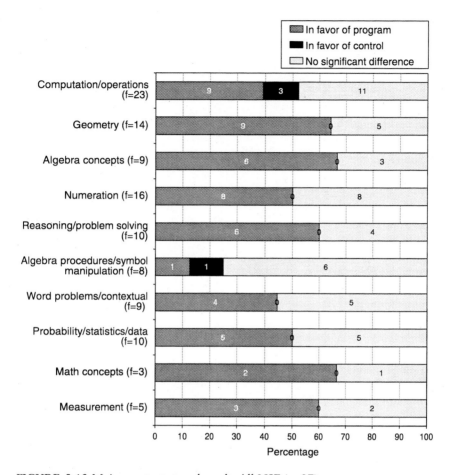

FIGURE 5-12 Major content strand result: All NSF (n=27).

According to these tentative results, future evaluations should examine whether the NSF-supported programs produce sufficient competency among students in the areas of algebraic manipulation and computation. In computation, approximately 40 percent of the results were in favor of the treatment group, no significant differences were reported in approximately 50 percent of the results, and results in favor of the comparison were revealed 10 percent of the time. Interpreting that final proportion of no significant difference is essential. Some would argue that because computation has not been emphasized, findings of no significant differences are acceptable. Others would suggest that such findings indicate weakness, because the development of the materials and accompanying professional development yielded no significant difference in key areas.

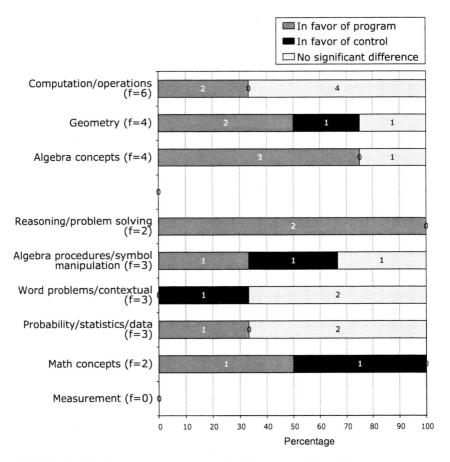

FIGURE 5-13 Major content strand result: All commercial (n=8).

From Figure 5-13 of findings from studies of commercially generated curricula, it appears that mixed results are commonly reported. Thus, in evaluations of commercial materials, lack of significant differences in computations/operations, word problems, and probability and statistics suggest that careful attention should be given to measuring these outcomes in future evaluations.

Overall, the grade band results for the NSF-supported programs—while consistent with the aggregated results—provide more detail. At the elementary level, evaluations of NSF-supported curricula (n=12) report better performance in mathematics concepts, geometry, and reasoning and problem solving, and some weaknesses in computation. No content strand analysis for commercially generated materials was possible. Evaluations (n=6) at middle grades of NSF-supported curricula showed strength in measurement, geometry, and probability and statistics and some weaknesses in computation. In the studies of commercial materials, evaluations (n=4) reported favorable results in reasoning and problem solving and some unfavorable results in algebraic procedures, contextual problems, and mathematics concepts. Finally, at the high school level, the evaluations (n=9) by content strand for the NSF-supported curricula showed strong favorable results in algebra concepts, reasoning/problem solving, word problems, probability and statistics, and measurement. Results in favor of the control were reported in 25 percent of the algebra procedures and 33 percent of computation measures.

For the studies of commercial materials (n=4), only the geometry results favor the control group 25 percent of the time, with 50 percent having favorable results. Algebra concepts, reasoning, and probability and statistics also produced favorable results.

Equity Analysis of Comparative Studies

When the goal of providing a standards-based curriculum to all students was proposed, most people could recognize its merits: the replacement of dull, repetitive, largely dead-end courses with courses that would lead all students to be able, if desired and earned, to pursue careers in mathematics-reliant fields. It was clear that the NSF-supported projects, a stated goal of which was to provide standards-based courses to all students, called for curricula that would address the problem of too few students persisting in the study of mathematics. For example, as stated in the NSF Request for Proposals (RFP):

> Rather than prematurely tracking students by curricular objectives, secondary school mathematics should provide for *all* students a common core of mainstream mathematics differentiated instructionally by level of abstraction and formalism, depth of treatment and pace (National Science Foundation, 1991, p. 1). In the elementary level solicitation, a similar statement on causes for all students was made (National Science Foundation, 1988, pp. 4-5).

> Some, but not enough attention has been paid to the education of students who fall below the average of the class. On the other hand, because the above average students sometimes do not receive a demanding education, it may be incorrectly assumed they are easy to teach (National Science Foundation, 1989, p. 2).

Likewise, with increasing numbers of students in urban schools, and increased demographic diversity, the challenges of equity are equally significant for commercial publishers, who feel increasing pressures to demonstrate the effectiveness of their products in various contexts.

The problem was clearly identified: poorer performance by certain subgroups of students (minorities—non-Asian, LEP students, sometimes females) and a resulting lack of representation of such groups in mathematics-reliant fields. In addition, a secondary problem was acknowledged: Highly talented American students were not being provided adequate challenge and stimulation in comparison with their international counterparts. We relied on the concept of equity in examining the evaluation. Equity was contrasted to equality, where one assumed all students should be treated exactly the same (Secada et al., 1995). Equity was defined as providing opportunities and eliminating barriers so that the membership in a subgroup does not subject one to undue and systematically diminished possibility of success in pursuing mathematical study. Appropriate treatment therefore varies according to the needs of and obstacles facing any subgroup.

Applying the principles of equity to evaluate the progress of curricular programs is a conceptually thorny challenge. What is challenging is how to evaluate curricular programs on their progress toward equity in meeting the needs of a diverse student body. Consider how the following questions provide one with a variety of perspectives on the effectiveness of curricular reform regarding equity:

• Does one expect all students to improve performance, thus raising the bar, but possibly not to decrease the gap between traditionally well-served and under-served students?

• Does one focus on reducing the gap and devote less attention to overall gains, thus closing the gap but possibly not raising the bar?

• Or, does one seek evidence that progress is made on both challenges—seeking progress for all students and arguably faster progress for those most at risk?

Evaluating each of the first two questions independently seems relatively straightforward. When one opts for a combination of these two, the potential for tensions between the two becomes more evident. For example, how can one differentiate between the case in which the gap is closed because talented students are being underchallenged from the case in which the gap is closed because the low-performing students improved their progress at an increased rate? Many believe that nearly all mathematics curricula in this country are insufficiently challenging and rigorous. Therefore achieving modest gains across all ability levels with evidence of accelerated progress by at-risk students may still be criticized for failure to stimulate the top performing student group adequately. Evaluating curricula with regard to this aspect therefore requires judgment and careful methodological attention.

Depending on one's view of equity, different implications for the collection of data follow. These considerations made examination of the quality of the evaluations as they treated questions of equity challenging for the committee members. Hence we spell out our assumptions as precisely as possible:

• Evaluation studies should include representative samples of student demographics, which may require particular attention to the inclusion of underrepresented minority students from lower socioeconomic groups, females, and special needs populations (LEP, learning disabled, gifted and talented students) in the samples. This may require one to solicit participation by particular schools or districts, rather than to follow the patterns of commercial implementation, which may lead to an unrepresentative sample in aggregate.

• Analysis of results should always consider the impact of the program on the entire spectrum of the sample to determine whether the overall gains are distributed fairly among differing student groups, and not achieved as improvements in the mean(s) of an identifiable subpopulation(s) alone.

• Analysis should examine whether any group of students is systematically less well served by curricular implementation, causing losses or weakening the rate of gains. For example, this could occur if one neglected the continued development of programs for gifted and talented students in mathematics in order to implement programs focused on improving access for underserved youth, or if one improved programs solely for one group of language learners, ignoring the needs of others, or if one's study systematically failed to report high attrition affecting rates of participation of success or failure.

• Analyses should examine whether gaps in scores between significantly disadvantaged or underperforming subgroups and advantaged subgroups are decreasing both in relation to eliminating the development of gaps in the first place and in relation to accelerating improvement for underserved youth relative to their advantaged peers at the upper grades.

In reviewing the outcomes of the studies, the committee reports first on what kinds of attention to these issues were apparent in the database, and second on what kinds of results were produced. Some of the studies used multiple methods to provide readers with information on these issues. In our report on the evaluations, we both provide descriptive information on the approaches used and summarize the results of those studies. Developing more effective methods to monitor the achievement of these objectives may need to go beyond what is reported in this study.

Among the 63 *at least minimally methodologically adequate* studies, 26 reported on the effects of their programs on subgroups of students. The

TABLE 5-13 Most Common Subgroups Used in the Analyses and the Number of Studies That Reported on That Variable

Identified Subgroup	Number of Studies of NSF-Supported	Number of Studies of Commercially Generated	Total
Gender	14	5	19
Race and ethnicity	14	2	16
Socioeconomic status	8	2	10
Achievement levels[a]	5	3	8
English as a second language (ESL)	2	1	3
Total	43	13	56

[a]Achievement levels: Outcome data are reported in relation to categorizations by quartiles or by achievement level based on independent test.

other 37 reported on the effects of the curricular intervention on means of whole groups and their standard deviations, but did not report on their data in terms of the impact on subpopulations. Of those 26 evaluations, 19 studies were on NSF-supported programs and 7 were on commercially generated materials. Table 5-13 reports the most common subgroups used in the analyses and the number of studies that reported on that variable. Because many studies used multiple categories for disaggregation (ethnicity, SES, and gender), the number of reports is more than double the number of studies. For this reason, we report the study results in terms of the "frequency of reports on a particular subgroup" and distinguish this from what we refer to as "study counts." The advantage of this approach is that it permits reporting on studies that investigated multiple ways to disaggregate their data. The disadvantage is that in a sense, studies undertaking multiple disaggregations become overrepresented in the data set as a result. A similar distinction and approach were used in our treatment of disaggregation by content strands.

It is apparent from these data that the evaluators of NSF-supported curricula documented more equity-based outcomes, as they reported 43 of the 56 comparisons. However, the same percentage of the NSF-supported evaluations disaggregated their results by subgroup, as did commercially generated evaluations (41 percent in both cases). This is an area where evaluations of curricula could benefit greatly from standardization of expectation and methodology. Given the importance of the topic of equity, it should be standard practice to include such analyses in evaluation studies.

In summarizing these 26 studies, the first consideration was whether representative samples of students were evaluated. As we have learned from medical studies, if conclusions on effectiveness are drawn without careful attention to representativeness of the sample relative to the whole population, then the generalizations drawn from the results can be seriously flawed. In Chapter 2 we reported that across the studies, approximately 81 percent of the comparative studies and 73 percent of the case studies reported data on school location (urban, suburban, rural, or state/region), with suburban students being the largest percentage in both study types. The proportions of students studied indicated a tendency to undersample urban and rural populations and oversample suburban schools. With a high concentration of minorities and lower SES students in these areas, there are some concerns about the representativeness of the work.

A second consideration was to see whether the achievement effects of curricular interventions were achieved evenly among the various subgroups. Studies answered this question in different ways. Most commonly, evaluators reported on the performance of various subgroups in the treatment conditions as compared to those same subgroups in the comparative condition. They reported outcome scores or gains from pretest to posttest. We refer to these as "between" comparisons.

Other studies reported on the differences among subgroups within an experimental treatment, describing how well one group does in comparison with another group. Again, these reports were done in relation either to outcome measures or to gains from pretest to posttest. Often these reports contained a time element, reporting on how the internal achievement patterns changed over time as a curricular program was used. We refer to these as "within" comparisons.

Some studies reported both between and within comparisons. Others did not report findings by comparing mean scores or gains, but rather created regression equations that predicted the outcomes and examined whether demographic characteristics are related to performance. Six studies (all on NSF-supported curricula) used this approach with variables related to subpopulations. Twelve studies used ANCOVA or Multiple Analysis of Variance (MANOVA) to study disaggregation by subgroup, and two reported on comparative effect sizes. In the studies using statistical tests other than t-tests or Chi-squares, two were evaluations of commercially generated materials and the rest were of NSF-supported materials.

Of the studies that reported on gender (n=19), the NSF-supported ones (n=13) reported five cases in which the females outperformed their counterparts in the controls and one case in which the female-male gap decreased within the experimental treatments across grades. In most cases, the studies present a mixed picture with some bright spots, with the majority showing no significant difference. One study reported significant improvements for African-American females.

In relation to race, 15 of 16 reports on African Americans showed positive effects in favor of the treatment group for NSF-supported curricula. Two studies reported decreases in the gaps between African Americans and whites or Asians. One of the two evaluations of African Americans, performance reported for the commercially generated materials, showed significant positive results, as mentioned previously.

For Hispanic students, 12 of 15 reports of the NSF-supported materials were significantly positive, with the other 3 showing no significant difference. One study reported a decrease in the gaps in favor of the experimental group. No evaluations of commercially generated materials were reported on Hispanic populations. Other reports on ethnic groups occurred too seldom to generalize.

Students from lower socioeconomic groups fared well, according to reported evaluations of NSF-supported materials (n=8), in that experimental groups outperformed control groups in all but one case. The one study of commercially generated materials that included SES as a variable reported no significant difference. For students with limited English proficiency, of the two evaluations of NSF-supported materials, one reported significantly more positive results for the experimental treatment. Likewise, one study of commercially generated materials yielded a positive result at the elementary level.

We also examined the data for ability differences and found reports by quartiles for a few evaluation studies. In these cases, the evaluations showed results across quartiles in favor of the NSF-supported materials. In one case using the same program, the lower quartiles showed the most improvement, and in the other, the gains were in the middle and upper groups for the Iowa Test of Basic Skills and evenly distributed for the informal assessment.

Summary Statements

After reviewing these studies, the committee observed that examining differences by gender, race, SES, and performance levels should be examined as a regular part of any review of effectiveness. We would recommend that all comparative studies report on both "between" and "within" comparisons so that the audience of an evaluation can simply and easily consider the level of improvement, its distribution across subgroups, and the impact of curricular implementation on any gaps in performance. Each of the major categories—gender, race/ethnicity, SES, and achievement level—contributes a significant and contrasting view of curricular impact. Furthermore, more sophisticated accounts would begin to permit, across studies, finer distinctions to emerge, such as the effect of a program on young African-American women or on first generation Asian students.

In addition, the committee encourages further study and deliberation on the use of more complex approaches to the examination of equity issues. This is particularly important due to the overlaps among these categories, where poverty can show itself as its own variable but also may be highly correlated to prior performance. Hence, the use of one variable can mask differences that should be more directly attributable to another. The committee recommends that a group of measurement and equity specialists confer on the most effective design to advance on these questions.

Finally, it is imperative that evaluation studies systematically include demographically representative student populations and distinguish evaluations that follow the commercial patterns of use from those that seek to establish effectiveness with a diverse student population. Along these lines, it is also important that studies report on the impact data on all substantial ethnic groups, including whites. Many studies, perhaps because whites were the majority population, failed to report on this ethnic group in their analyses. As we saw in one study, where Asian students were from poor homes and first generation, any subgroup can be an at-risk population in some setting, and because gains in means may not necessarily be assumed to translate to gains for all subgroups or necessarily for the majority subgroup. More complete and thorough descriptions and configurations of characteristics of the subgroups being served at any location—with careful attention to interactions—is needed in evaluations.

Interactions Among Content and Equity, by Grade Band

By examining disaggregation by content strand by grade levels, along with disaggregation by diverse subpopulations, the committee began to discover grade band patterns of performance that should be useful in the conduct of future evaluations. Examining each of these issues in isolation can mask some of the overall effects of curricular use. Two examples of such analysis are provided. The first example examines all the evaluations of NSF-supported curricula from the elementary level. The second examines the set of evaluations of NSF-supported curricula at the high school level, and cannot be carried out on evaluations of commercially generated programs because they lack disaggregation by student subgroup.

Example One

At the elementary level, the findings of the review of evaluations of data on effectiveness of NSF-supported curricula report consistent patterns of benefits to students. Across the studies, it appears that positive results are enhanced when accompanied by adequate professional development and the use of pedagogical methods consistent with those indicated by the curricula. The benefits are most consistently evidenced in the broadening topics of geometry, measurement, probability, and statistics, and in applied problem solving and reasoning. It is important to consider whether the outcome measures in these areas demonstrate a depth of understanding. In early understanding of fractions and algebra, there is some evidence of improvement. Weaknesses are sometimes reported in the areas of computational skills, especially in the routinization of multiplication and division. These assertions are tentative due to the possible flaws in designs but quite consistent across studies, and future evaluations should seek to replicate, modify, or discredit these results.

The way to most efficiently and effectively link informal reasoning and formal algorithms and procedures is an open question. Further research is needed to determine how to most effectively link the gains and flexibility associated with student-generated reasoning to the automaticity and generalizability often associated with mastery of standard algorithms.

The data from these evaluations at the elementary level generally present credible evidence of increased success in engaging minority students and students in poverty based on reported gains that are modestly higher for these students than for the comparative groups. What is less well documented in the studies is the extent to which the curricula counteract the tendencies to see gaps emerge and result in long-term persistence in performance by gender and minority group membership as they move up the grades. However, the evaluations do indicate that these curricula can help, and almost never do harm. Finally, on the question of adequate challenge for advanced and talented students, the data are equivocal. More attention to this issue is needed.

Example Two

The data at the high school level produced the most conflicting results, and in conducting future evaluations, evaluators will need to examine this level more closely. We identify the high school as the crucible for curricular change for three reasons: (1) the transition to postsecondary education puts considerable pressure on these curricula; (2) the criteria outlined in the NSF RFP specify significant changes from traditional practice; and (3) high school freshmen arrive from a myriad of middle school curricular experiences. For the NSF-supported curricula, the RFP required that the programs provide a core curriculum "drawn from statistics/probability, algebra/functions, geometry/trigonometry, and discrete mathematics" (NSF, 1991, p. 2) and use "a full range of tools, including graphing calculators and computers" (NSF, 1991, p. 2). The NSF RFP also specified the inclusion of "situations from the natural and social sciences and from other parts of the school curriculum as contexts for developing and using mathematics" (NSF, 1991, p. 1). It was during the fourth year that "course options should focus on special mathematical needs of individual students, accommodating not only the curricular demands of the college-bound but also specialized applications supportive of the workplace aspirations of employment-bound students" (NSF, 1991, p. 2). Because this set of requirements comprises a significant departure from conventional practice, the implementation of the high school curricula should be studied in particular detail.

We report on a Systemic Initiative for Montana Mathematics and Science (SIMMS) study by Souhrada (2001) and Brown et al. (1990), in which students were permitted to select traditional, reform, and mixed tracks. It became apparent that the students were quite aware of the choices they faced, as illustrated in the following quote:

> The advantage of the traditional courses is that you learn—just math. It's not applied. You get a lot of math. You may not know where to use it, but you learn a lot. . . . An advantage in SIMMS is that the kids in SIMMS tell me that they really understand the math. They understand where it comes from and where it is used.

This quote succinctly captures the tensions reported as experienced by students. It suggests that student perceptions are an important source of evidence in conducting evaluations. As we examined these curricular evaluations across the grades, we paid particular attention to the specificity of the outcome measures in relation to curricular objectives. Overall, a review of these studies would lead one to draw the following tentative summary conclusions:

- There is some evidence of discontinuity in the articulation between high school and college, resulting from the organization and emphasis of the new curricula. This discontinuity can emerge in scores on college admission tests, placement tests, and first semester grades where nonreform students have shown some advantage on typical college achievement measures.
- The most significant areas of disadvantage seem to be in students' facility with algebraic manipulation, and with formalization, mathematical structure, and proof when isolated from context and denied technological supports. There is some evidence of weakness in computation and numeration, perhaps due to reliance on calculators and varied policies regarding their use at colleges (Kahan, 1999; Huntley et al., 2000).
- There is also consistent evidence that the new curricula present

strengths in areas of solving applied problems, the use of technology, new areas of content development such as probability and statistics and functions-based reasoning in the use of graphs, using data in tables, and producing equations to describe situations (Huntley et al., 2000; Hirsch and Schoen, 2002).

• Despite early performance on standard outcome measures at the high school level showing equivalent or better performance by reform students (Austin et al., 1997; Merlino and Wolff, 2001), the common standardized outcome measures (Preliminary Scholastic Assessment Test [PSAT] scores or national tests) are too imprecise to determine with more specificity the comparisons between the NSF-supported and comparison approaches, while program-generated measures lack evidence of external validity and objectivity. There is an urgent need for a set of measures that would provide detailed information on specific concepts and conceptual development over time and may require use as embedded as well as summative assessment tools to provide precise enough data on curricular effectiveness.

• The data also report some progress in strengthening the performance of underrepresented groups in mathematics relative to their counterparts in the comparative programs (Schoen et al., 1998; Hirsch and Schoen, 2002).

This reported pattern of results should be viewed as very tentative, as there are only a few studies in each of these areas, and most do not adequately control for competing factors, such as the nature of the course received in college. Difficulties in the transition may also be the result of a lack of alignment of measures, especially as placement exams often emphasize algebraic proficiencies. These results are presented only for the purpose of stimulating further evaluation efforts. They further emphasize the need to be certain that such designs examine the level of mathematical reasoning of students, particularly in relation to their knowledge of understanding of the role of proofs and definitions and their facility with algebraic manipulation as we as carefully document the competencies taught in the curricular materials. In our framework, gauging the ease of transition to college study is an issue of examining curricular alignment with systemic factors, and needs to be considered along with those tests that demonstrate a curricular validity of measures. Furthermore, the results raising concerns about college success need replication before secure conclusions are drawn.

Also, it is important that subsequent evaluations also examine curricular effects on students' interest in mathematics and willingness to persist in its study. Walker (1999) reported that there may be some systematic differences in these behaviors among different curricula and that interest and persistence may help students across a variety of subgroups to survive entry-level hurdles, especially if technical facility with symbol manipulation can be improved. In the context of declines in advanced study in mathematics by American students (Hawkins, 2003), evaluation of curricular impact on students' interest, beliefs, persistence, and success are needed.

The committee takes the position that ultimately the question of the impact of different curricula on performance at the collegiate level should be resolved by whether students are adequately prepared to pursue careers in mathematical sciences, broadly defined, and to reason quantitatively about societal and technological issues. It would be a mistake to focus evaluation efforts solely or primarily on performance on entry-level courses, which can clearly function as filters and may overly emphasize procedural competence, but do not necessarily represent what concepts and skills lead to excellence and success in the field.

These tentative patterns of findings indicate that at the high school level, it is necessary to conduct individual evaluations that examine the transition to college carefully in order to gauge the level of success in preparing students for college entry and the successful negotiation of majors. Equally, it is imperative to examine the impact of high school curricula on other possible student trajectories, such as obtaining high school diplomas, moving into worlds of work or through transitional programs leading to technical training, two-year colleges, and so on.

These two analyses of programs by grade-level band, content strand, and equity represent a methodological innovation that could strengthen the empirical database on curricula significantly and provide the level of detail really needed by curriculum designers to improve their programs. In addition, it appears that one could characterize the NSF programs (and not the commercial programs as a group) as representing a particular approach to curriculum, as discussed in Chapter 3. It is an approach that integrates content strands; relies heavily on the use of situations, applications, and modeling; encourages the use of technology; and has a significant dose of mathematical inquiry. One could ask the question of whether this approach as a whole is "effective." It is beyond the charge and scope of this report, but is a worthy target of investigation if one uses proper care in design, execution, and analysis. Likewise other approaches to curricular change should be investigated at the aggregate level, using careful and rigorous design.

The committee believes that a diversity of curricular approaches is a strength in an educational system that maintains local and state control of curricular decision making. While "scientifically established as effective" should be an increasingly important consideration in curricular choice, local cultural differences, needs, values, and goals will also properly influence curricular choice. A diverse set of effective curricula would be ideal. Finally, the committee emphasizes once again the importance of basing the studies on measures with established curricular validity and avoiding corruption of indicators as a result of inappropriate amounts of teaching to the test, so as to be certain that the outcomes are the product of genuine student learning.

CONCLUSIONS FROM THE COMPARATIVE STUDIES

In summary, the committee reviewed a total of 95 comparative studies. There were more NSF-supported program evaluations than commercial ones, and the commercial ones were primarily on Saxon or UCSMP materials. Of the 19 curricular programs reviewed, 23 percent of the NSF-supported and 33 percent of the commercially generated materials selected had programs with no comparative reviews. This finding is particularly disturbing in light of the legislative mandate in No Child Left Behind (U.S. Department of Education, 2001) for scientifically based curricular programs and materials to be used in the schools. It suggests that more explicit protocols for the conduct of evaluation of programs that include comparative studies need to be required and utilized.

Sixty-nine percent of NSF-supported and 61 percent of commercially generated program evaluations met basic conditions to be classified as *at least minimally methodologically adequate* studies for the evaluation of effectiveness. These studies were ones that met the criteria of including measures of student outcomes on mathematical achievement, reporting a method of establishing comparability among samples and reporting on implementation elements, disaggregating by content strand, or using precise, theoretical analyses of the construct or multiple measures.

Most of these studies had both strengths and weaknesses in their quasi-experimental designs. The committee reviewed the studies and found that evaluators had developed a number of features that merit inclusions in future work. At the same time, many had internal threats to validity that suggest a need for clearer guidelines for the conduct of comparative evaluations.

Many of the strengths and innovations came from the evaluators' understanding of the program theories behind the curricula, their knowledge of the complexity of practice, and their commitment to measuring valid and significant mathematical ideas. Many of the weaknesses came from inadequate attention to experimental design, insufficient evidence of the independence of evaluators in some studies, and instability and lack of cooperation in interfacing with the conditions of everyday practice.

The committee identified 10 elements of comparative studies needed to establish a basis for determining the effectiveness of a curriculum. We recognize that not all studies will be able to implement successfully all elements, and those experimental design variations will be based largely on study size and location. The list of elements begins with the seven elements corresponding to the seven critical decisions and adds three additional elements that emerged as a result of our review:

1. A better balance needs to be achieved between experimental and quasi-experimental studies. The virtual absence of large-scale experimental studies does not provide a way to determine whether the use of quasi-experimental approaches is being systematically biased in unseen ways.

2. If a quasi-experimental design is selected, it is necessary to establish comparability. When quasi-experimentation is used, it "pertains to studies in which the model to describe effects of secondary variables is not known but assumed" (NRC, 1992, p. 18). This will lead to weaker and potentially suspect causal claims, which should be acknowledged in the evaluation report, but may be necessary in relation to feasibility (Joint Committee on Standards for Educational Evaluation, 1994). In general, to date, studies have assumed prior achievement measures, ethnicity, gender, and SES, are acceptable variables on which to match samples or on which to make statistical adjustments. But there are often other variables in need of such control in such evaluations including opportunity to learn, teacher effectiveness, and implementation (see #4 below).

3. The selection of a unit of analysis is of critical importance to the design. To the extent possible, it is useful to randomly assign the unit for the different curricula. The number of units of analysis necessary for the study to establish statistical significance depends not on the number of students, but on this unit of analysis. It appears that classrooms and schools are the most likely units of analysis. In addition, the development of increasingly sophisticated means of conducting studies that recognize that the level of the educational system in which experimentation occurs affects research designs.

4. It is essential to examine the implementation components through a set of variables that include the extent to which the materials are implemented, teaching methods, the use of supplemental materials, professional development resources, teacher background variables, and teacher effects. Gathering these data to gauge the level of implementation fidelity is essential for evaluators to ensure adequate implementation. Studies could also include nested designs to support analysis of variation by implementation components.

5. Outcome data should include a variety of measures of the highest quality. These measures should vary by question type (open ended, multiple choice), by type of test (international, national, local) and by relation of testing to everyday practice (formative, summative, high stakes), and ensure curricular validity of measures and assess curricular alignment with systemic factors. The use of comparisons among total tests, fair tests, and

conservative tests, as done in the evaluations of UCSMP, permits one to gain insight into teacher effects and to contrast test results by items included. Tests should also include content strands to aid disaggregation, at a level of major content strands (see Figure 5-11) and content-specific items relevant to the experimental curricula.

6.   Statistical analysis should be conducted on the appropriate unit of analysis and should include more sophisticated methods of analysis such as ANOVA, ANCOVA, MACOVA, linear regression, and multiple regression analysis as appropriate.

7.   Reports should include clear statements of the limitations to generalization of the study. These should include indications of limitations in populations sampled, sample size, unique population inclusions or exclusions, and levels of use or attrition. Data should also be disaggregated by gender, race/ethnicity, SES, and performance levels to permit readers to see comparative gains across subgroups both between and within studies.

8.   It is useful to report effect sizes. It is also useful to present item-level data across treatment program and show when performances between the two groups are within the 10 percent confidence interval of each other. These two extremes document how crucial it is for curricula developers to garner both precise and generalizable information to inform their revisions.

9.   Careful attention should also be given to the selection of samples of populations for participation. These samples should be representative of the populations to whom one wants to generalize the results. Studies should be clear if they are generalizing to groups who have already selected the materials (prior users) or to populations who might be interested in using the materials (demographically representative).

10.   The control group should use an identified comparative curriculum or curricula to avoid comparisons to unstructured instruction.

In addition to these prototypical decisions to be made in the conduct of comparative studies, the committee suggests that it would be ideal for future studies to consider some of the overall effects of these curricula and to test more directly and rigorously some of the findings and alternative hypotheses. Toward this end, the committee reported the tentative findings of these studies by program type. Although these results are subject to revision, based on the potential weaknesses in design of many of the studies summarized, the form of analysis demonstrated in this chapter provides clear guidance about the kinds of knowledge claims and the level of detail that we need to be able to judge effectiveness. Until we are able to achieve an array of comparative studies that provide valid and reliable information on these issues, we will be vulnerable to decision making based excessively on opinion, limited experience, and preconceptions.

6

Case Studies and Synthesis Studies

CASE STUDIES

The committee drew a distinction between two types of studies: comparative studies that investigate a curriculum's effectiveness (as indicated by measures of student performance) and case studies, which typically examine the mechanism or means of obtaining those effects, although some case studies do attend to outcome measures. Case studies typically document "what happened" differently than do comparative studies. Case studies provide insight into mechanisms at play that are hidden from a comparison of student achievement. This is an important distinction for program evaluation and curriculum development, as the actual treatment in a large-scale comparative study is often ill defined. As discussed in a report by the National Research Council (2002, p. 117):

> In many situations, finding that a causal agent (x) leads to the outcome (y) is not sufficient. Important questions remain about *how x* causes *y*. Questions about how things work demand attention to the processes and mechanisms by which the causes produced their effects.

Case study research is appropriate "when the inquirer seeks answers to 'how' or 'why' questions, when the inquirer has little control over events being studied, when the object of study is a contemporary phenomenon in a real-life context, when boundaries between the phenomenon and the context are not clear, and when it is desirable to use multiple sources of evidence" (Schwandt, 2001, p. 23).

Although the genre of studies that investigate details of "what happened" frequently do not provide sufficient experimental evidence to permit causal inference about a curriculum's effectiveness as measured by student achievement, the studies may indicate why that curriculum had the effect it did and it may highlight aspects of implementation or design that were instrumental in producing that effect. Such studies, generally referred to as case studies, can provide useful information along a number of dimensions that emanate from a careful description of the connections among a curriculum's program theory, its implementation theory, and its actualization in particular settings (Bickman, 1987). The generalizations from a well-designed comparative evaluation may not provide sufficient information to permit decision makers to know whether the experimental treatment (new curriculum) will be appropriate for their particular setting. Case studies may provide additional specificity that is necessary and helpful to practitioners in assessing the probability of successful use in their settings. As written by Easley (1977, p. 6):

> Experimentalists feel that they can generalize their findings from an experiment to the population as a whole because they have drawn an adequate random sample from the population about which a hypothesis speaks. Clinical researchers feel that they can generalize from a study of a single case to some other individual cases because they have seen a given phenomenon in one situation in sufficient detail and know its essential workings to be able to recognize it when they encounter it in another situation.

Criteria for Inclusion in Our Study

Forty-five articles, dissertations, and unpublished manuscripts were originally classified as case studies. We considered case studies, ethnographies, descriptive studies, and research studies that inform us about what happens in the implementation of specific curricula, classifying them all as "case studies" for simplicity. To be classified as a case study, the study had to examine curricula implementation of significant parts of the curricula materials (more than one unit) over a significant duration (more than one semester) and had to show evidence of systematic data collection and report on the effectiveness of the materials in the conclusions. For our purposes the study also had to focus on 1 of the 13 mathematics curricula supported by the National Science Foundation (NSF), the University of Chicago School Mathematics Project (UCSMP) curriculum, or one of the five other commercially generated mathematics curricula included in our review.

After the initial categorizing, we refined our criteria for inclusion in our review to stipulate that the case studies must have been published, be a dissertation, or have a draft date of 2000 or later. We assumed that manuscripts with a draft date prior to 2000 were written with the intent to publish. Therefore we decided not to consider them if they remained unpublished in 2003. Unpublished manuscripts written prior to 2000 will probably not be published even if they were written with that intention. On the other hand, manuscripts with draft dates of 2000 or later may be in the pipeline for publication and were included. Thirty-two studies met our criteria.

The Studies

From the original 45 studies, we excluded 12 draft manuscripts with dates prior to 2000 and 1 manuscript dated in 2000 because it was simply a compilation of the author's dissertation results. Thus we included 32 studies: 9 unpublished manuscripts, 13 dissertations, and 10 published articles. Therefore, the remainder of the section will report only on the 32 included studies. Table 6-1 reports the number of case studies on each NSF-supported curriculum.

TABLE 6-1 Distribution of Case Studies by Curricula

	Number of Studies
NSF-Supported Elementary Curriculum Materials	
Everyday Mathematics (EM)	4
Investigations in Number, Data and Space/TERC	1
Math Trailblazers	1
NSF-Supported Middle School Curriculum Materials	
Connected Mathematics Project (CMP)	14
Mathematics in Context (MiC)	7
*Math*Thematics *(STEM)*	4
MathScape	2
MS Mathematics Through Applications Project (MMAP)	0
NSF-Supported High School Curriculum Materials	
Interactive Mathematics Program (IMP)	1
Mathematics: Modeling Our World (MMOW or ARISE)	0
Contemporary Mathematics in Context (Core-Plus)	5
Math Connections	0
SIMMS	1
Commercially Generated Elementary Curriculum Materials	
Addison Wesley: Math, 2002	0
Harcourt Brace: Harcourt Math K-6	0
Commercially Generated Middle School Curriculum Materials	
McGraw-Hill/Glencoe: Applications and Connections, 2001	0
Saxon: An Incremental Development	0
Commercially Generated High School Curriculum Materials	
Houghton Mifflin/McDougal Littell: Larson Series, 2002	0
Prentice Hall: UCSMP Integrated Mathematics, 2002	0

NOTE: Some reports addressed more than one curriculum, so the number of curricula addressed is larger than the number of included studies.

Method

We judged each included study according to how well it met the following criteria:

1. *Defined the case.*

A report defined its case well if it made clear the category to which the case belonged. In other words, a well-defined case allowed us to make statements that clarified, "This study is about x," where x was defined with enough specificity and clarity that an equivalent case could be replicated at a later time with assurance of studying a similar phenomenon. "This study is about two middle school teachers teaching a reform curriculum" is not sufficiently clear about the subjects or the setting to assure an equivalent case in another study. In terms of our framework, a well-defined case also presented a clearly articulated program theory.

2. *Backed its claims by evidence and argument.*

Authors backed their claims when they used a methodology that included data, a way to analyze data systematically, and a form of argument that could support a reader's reaching a similar or contrary conclusion. This criterion permitted us to distinguish a case study from an anecdotal report that told a story, but did not indicate how the data were systematically collected, linked to program theory, and analyzed and evaluated.

3. *Was based on a replicable design.*

A central feature of a scientific experiment is that the conditions under which it was conducted, the procedures used in conducting it, and the methods for collecting and analyzing data are described explicitly. The purpose of explicit and veridical descriptions is so that other researchers can perform "the same" experiment or a variation of it in order to compare its results with the experiment being replicated. Thus, replicability of an experiment does not refer to the experiment's results being repeated. Rather, it refers to repeating the experiment itself so that the replication's results can be compared with the original's results. Clearly a case cannot be precisely repeated. But the method of constructing a case can be repeated if conducted appropriately and if described sufficiently, and if not repeated precisely the differences can be noted and taken into account. Thus, a case must provide sufficient delineation of case events, behaviors, perceptions, and the methods of data collection associated with them to permit another evaluator to design a parallel evaluation in another setting and to conduct a related study. A replicable design, therefore, is one that allows another person to "repeat" the study methodologically, to the extent feasible, using similar data collection techniques and similar analytic methods.

4. *Revealed something about the mechanisms at play during the implementation of a curriculum.*

A case study should develop clear explanatory constructs that coherently link together the mechanisms involved in curricular use with the program theory, the conditions of implementation, and the documented events, behaviors, and perceptions of the case.

The studies included in our review are most valuable for generating explanations about a curriculum's program theory or implementation theory. This stance is in line with Campbell's (1994) thinking that generating and addressing rival explanations of a phenomenon is the heart of scientific inquiry. As pointed out in the section on comparative studies, typical experimental and quasi-experimental designs produce results subject to refutation by rival hypotheses. Case studies can be useful in shedding light on which of these hypotheses, or others, are most promising to pursue.

Each included study was read by at least two committee members and discussed with regard to each criterion, thereby leading to a consensus score of 1 (poor), 2 (acceptable), or 3 (well done) being assigned on each.

Findings

Case studies in this review were found only in the NSF-supported mathematics curriculum materials. Therefore, the generalization of results must be restricted to NSF-supported curricula; case studies of commercially generated curricula would be needed to draw broader conclusions. Table 6-2 provides an overview of how many studies received what rating on each of the four criteria.

Only 11 studies received ratings of 2 or 3 on all criteria. Surprisingly, there was little correlation between quality rankings and type of report. Table 6-3 shows the breakdown of level by type of report. Dissertations tended to back claims better than articles or manuscripts. Dissertations and articles tended to have higher ranks than unpublished manuscripts, which were notably poor at providing a sense of mechanism by which a curriculum's effects might be realized. However, on any criterion the majority of reports were at best acceptable.

TABLE 6-2 Number of Studies by Rating on Each Criterion

Quality Ranking	Defined Case	Backed Claims	Replicable Design	Insight Into Mechanism
Level 1 (poor)	16	15	16	17
Level 2 (acceptable)	7	5	6	10
Level 3 (well done)	9	12	10	5

Patterns in Findings

Despite the relatively small number of high-quality studies, there were four recurrent issues that were raised broadly in these studies and bear on the design of future evaluations. It is important to note that our purpose for identifying patterns in case study results is methodological. Case studies can provide useful information on how program components interact with implementation factors at the level of classroom practices, and therefore can provide insight into the reason for whatever level of curricular effectiveness occurred. Case studies can therefore inform future evaluators about potential explanatory variables to include the conduct of future evaluations.

Recurrent issues among the case studies were as follows:

• Design features affect student subpopulations differentially;
• Common practices, beliefs, and understandings among teachers and students interact in unanticipated ways with characteristics of these curricula;
• Professional development is an essential consideration; and
• Time and resource allocations must be carefully managed.

A fifth issue that ultimately could be important for curriculum adopters, and thus potentially important to be addressed in evaluations, is that of students' transitions from reform to nonreform curricula (or vice versa). One study (de Groot, 2000) followed three female students as they transitioned from the Connected Mathematics Project (CMP) to standard 9th-grade algebra, identifying interesting issues that might generalize to the larger population. However, there were no other studies of sufficient quality addressing this issue to report any secure or robust patterns.

Differential Impact on Different Student Populations

Many new curricula anticipate that instruction will be highly interactive, involving students in patterns resembling reflective discourse in which students and teacher interact around substantive ideas and take their under-

TABLE 6-3 Type of Studies by Rating on Each Criterion

Criterion Rating	Defined Case			Backed Claims			Replicable Design			Insight into Mechanism		
	1	2	3	1	2	3	1	2	3	1	2	3
Unpublished manuscript	6	3	0	7	1	1	7	0	2	8	0	1
Dissertation	6	2	5	3	2	8	4	4	5	5	6	2
Published article	4	2	4	5	2	3	5	2	3	4	4	2

NOTE: Criterion Rating: (1) poor; (2) acceptable; (3) well done.

standing of mathematics and their forms of representing it as objects of discussion (Cobb et al., 1997). Although it has been documented that these practices can offer advantages to students who participate in them (Nicholls et al., 1990; Cobb et al., 1991; Lehrer et al., 1999), it is possible that these practices are not easily implemented widely without greater attention to changes in classroom culture and teachers' expectations of why these practices might be fruitful.

Baxter et al. (2001) and Murphy (1998) suggest the importance of giving special attention to low-achieving students when implementing instructional practices that emphasize public displays of knowledge, such as working in small groups or participating in whole-class discussions. Woodward and Baxter (1997), in a comparison of Everyday Mathematics and Heath Mathematics, found that while average- and high-ability students seemed to benefit from using Everyday Mathematics in relation to the comparison groups, low-achieving students in both groups performed at comparatively the same level and showed only modest improvement over time. In a follow-up qualitative study of why low-achieving students benefited less than higher achieving students in Everyday Mathematics, Baxter et al. (2001) found that low-achieving students often were disengaged during whole-class discussion. Sometimes they were not able to follow other students' often poorly constructed and fragmentary contributions. At other times the nature of the discussion seemed to assume levels of prior knowledge that many low-ability students lacked.

An exception to this finding occurred during small-group work. Low-achieving students were more engaged during small-group work than in whole-class instruction. However, the nature of their engagement was typically low level (e.g., copying results, collecting resources). Baxter et al. (2001) noted that one teacher was successful using Everyday Mathematics with low-achieving students. The difference was that this teacher provided many conceptual entry points to conversations, a point elaborated further in the next section. Murphy (1998) suggested an additional reason why low-achieving students failed to participate in discussions by noting that low-achieving students felt greater exposure to ridicule because they had to display their lack of understanding or achievement in front of others.

The Baxter et al. and Murphy studies do not provide comparative assessments of curricula that place a premium on group work or class discussion. Rather, they are most useful in generating hypotheses about what kinds of practices associated with such curricula may need to be modified or supplemented to ensure a fair distribution of opportunities to learn among all ability levels. They suggest a need for curricular evaluations to examine whether implementation and program theories provide sufficient attention to necessary changes in existing classroom norms and practices for various subgroups, and to study relationships between actualization of those theories and student achievement.

In a dissertation focusing on the CMP curriculum, Lubienski (2000) argued that differences in socio-economic status (SES) among students can be linked to clashes between curriculum designers' intent to empower students mathematically and cultural values internalized by them. In particular, she focuses on the responses of low-SES students to the context of open-ended, ill-defined problems:

> Hence, in contrast with the reformers' rhetoric of "mathematical empowerment," some of my students reacted to the more open, challenging mathematics problems by becoming overly frustrated and feeling increasingly mathematically disempowered. The lower SES students seemed to prefer more external direction from the textbook and the teacher. The lower SES students, particularly the females, seemed to internalize their struggles and "shut down," preferring a more traditional, directive role from the teacher and text. These students longed to return to the days in which they could see more direct results for their efforts (e.g., 48 out of 50 correct on the day's worksheet). (p. 476)

Lubienski emphasized that readers should not generalize her observations to all implementations of problem-based mathematics curricula. Instead she stressed that the strongest use of her results should be to alert designers and users of problem-centered curricula of the possibility that they may be insensitive to cultural values designed into curricula that certain student subpopulations may not initially or subsequently understand or share. We add to Lubienski's caveat that evaluations should be designed with the awareness of possible unintended interactions between program design and subgroup characteristics.

In a similar vein, Hetherington (2000) documented a clash between the emphasis that Core-Plus Mathematics places on group work and public discourse and the habitual lack of intellectual engagement that students in her study had developed in prior years. Late, sloppy work interfered with progress because curriculum designers anticipated that later tasks and assignments would build on previous, solid work that students had not accomplished.

Taken together, these examples illustrate the complex interactions among key features of a program's design, existing instructional practices, and characteristics of particular student subgroups that program evaluations should consider. Attending to these interactions in program evaluations may provide more precise understandings of a curriculum's differential impact among student subgroups and of its differential impact among implementation sites. It might also lead to a deeper understanding about how cognitive and conceptual accomplishments are produced through the interactions among curricular tasks and student and teacher participation patterns (Greeno and Goldman, 1998).

Interactions Among Curricula and Common Practices, Beliefs, and Understandings

The prior section focused on design features that assume students benefit automatically from public discussion as a means of support for collective reflection and student engagement. Those studies elucidate unexpected interactions between public discourse and student characteristics that can result in their disengagement rather than engagement. Complicating the framework further, several studies also document that these design characteristics can interact with teacher characteristics in ways that diminish curricular effects. Teachers who express mathematical ideas primarily in terms of numbers, symbols, and operations, and who encourage students to do the same, can create ways of talking that make the ideas being discussed accessible only to those who already understand the ideas—and therefore inaccessible to students who do not already understand them (Thompson et al., 1994).

Several studies (Fuson et al., no date; Herbel-Eisenmann, 2000; Manouchehri and Goodman, 1998, 2000) provided excerpts of classroom dialog suggesting that the degree to which teachers and students speak calculationally could be an important factor in how successfully they implement curricula that place a premium on public discourse in the service of teaching for understanding. Other studies suggested that the degree to which teachers are oriented to making sense of mathematical ideas for students can be important factors both in using public discussion productively (Kett, 1997; Smith, 1998) and in implementing the curriculum according to its designers' intent (Manouchehri and Goodman, 1998, 2000).

We found one study particularly informative in illustrating the evaluators' view of the importance of classroom discourse that draws on students' ideas. Fuson et al. (no date) analyzed 1st-grade Everyday Mathematics materials to discern the social and sociomathematical norms (Yackel and Cobb, 1996) assumed by the curriculum designers. Three particularly important norms were:

1. Extend students' thinking;
2. Use errors as opportunities for learning; and
3. Foster student-to-student discussion of mathematical thinking.

They investigated the degree to which these norms were implemented in 19 1st-grade classrooms in the Chicago area. In only one of the classrooms did the authors witness all three norms addressed. In attempting to understand why so few teachers implemented "extended students' thinking," they deduced that teachers needed to shift from talking about "their" and "the text's" mathematics to talking about children's mathematics.

In summary, studies highlighted in this section point to potentially significant sources of variation in the impact of NSF-supported mathematics curricula that should be addressed in program evaluations. A curriculum's program theory may presume a certain instructional discourse style that requires significant changes in teachers' beliefs and practices. It may also be designed with the anticipation that teachers will foster certain social and sociomathematical norms when it may be uncommon that they do.

Professional Development

The most compelling pattern among included case studies was the importance of professional development. Schoen et al. (2003) found that teachers' engagement in professional development, comfort with class management, and high performance expectations for their students were the best predictors of student achievement among a sample of teachers implementing the Core-Plus Mathematics Project. Collins (2002), in a comparative case study of one implementation of the Connected Mathematics Project, examined student achievement in relation to the level of teacher professional development in three Boston schools. Collins found that "students in schools whose teachers received sustained professional development designed to meet the needs of the participating teachers performed significantly higher on both the Massachusetts Comprehensive Assessment System (MCAS) and a nationally normed achievement test, TerraNova, than did those students whose teachers had not participated in consistent professional development" (p. 8).

Bay (1999) studied the effects of teacher collaboration on curricular implementation and determined that a lack of collaboration among teachers at an implementation site appeared to allow room for individual teachers' frustrations to foment, sometimes leading to their return to old routines. On the other hand, collaboration among teachers at an implementation site appeared to sustain excitement and commitment to change. Dapples (1994) found that teachers who implemented the Systemic Initiative for Montana Mathematics and Science (SIMMS) curriculum found professional development instrumental in implementing SIMMS. However, these same teachers were also teaching "traditional" courses and found few entry points to use their new routines learned in professional development.

These case studies show clearly that the level and quality of professional development entailed in a curriculum implementation are important factors in its effectiveness, especially when the curriculum demands changes in teachers' beliefs, understandings, and practices. Therefore, evaluators should document and measure the types of opportunities for teachers to participate in professional development and the frequency of these opportunities as well as the opportunities for teachers to collaborate (i.e., on curricular decision making and curricular implementation issues).

Time Management

While many studies pointed to the disruption in well-established routines that teachers had developed with other curricula, two studies in particular suggested processes that might systematically create time management problems. Hetherington (2000) provided a rich description of her mathematics department's earnest and well-intentioned attempt to implement Core-Plus Mathematics Project materials. Teachers experienced significant problems in managing the flow of instruction in the context of Core-Plus' emphasis on group work and, as already mentioned, students' predilections to be minimally engaged with instruction. Keiser and Lambdin (2001) documented the difficulty teachers had in organizing their instruction into coherent chunks that had educationally appropriate beginning and ending points and yet fit into fixed time blocks in the school day. They pointed out the difficulty of parsing a curriculum organized around conceptual themes into predetermined time blocks in comparison to doing the same with a curriculum that is organized by topics, facts, and procedures that typically are presented in smaller units.

Kramer and Keller (2003) suggested that a conceptually organized curriculum that is used in the schools with block scheduling can work better than a procedurally organized curriculum used in schools with traditional scheduling. In a finding reminiscent of the prior professional development section, Kett (1997) found that teachers had persistent difficulty in implementing new assessment procedures that involved a greater volume of student work, and student work also came in forms that placed higher demands on teachers' abilities to interpret the student's work and in-class contributions.

Comments on Case Studies Evaluations

It is worthwhile to note that case studies often reveal aspects of program components, implementation components, and their interactions that work differently than intended by program designers. This is one reason why case studies are a valuable tool in an evaluator's methodological toolkit. We note again that our sample of case studies was limited to studies of NSF-supported curricula and hence no broader generalizations can be drawn.

Although the case studies were valuable in pointing to important variables that should be included in future curriculum evaluations, the committee noted several aspects of the case studies as a group that deserve comment. Overall, the case studies displayed an inattention to theory and a disconnection with other research on learning and teaching mathematics, thereby limiting the ability of results to become cumulative across studies. The majority of studies told a story instead of developing a theory. At times, it seemed as if the documentation of contextual detail overpowered the building of theoretical, testable, and generalizable constructs and hence limited the potential to use these studies to contribute to an aggregated knowledge base on effectiveness.

Many studies would have been strengthened considerably if they had attempted to explain their observations by drawing on pertinent theories in the creation of constructs that pointed to mechanisms presented in the case that may be in play generally. Explaining observations requires systematic data collection and analysis, and it requires the investigator to entertain competing interpretations of what happened and competing explanations of why things happened as described. Often studies could have been strengthened by some degree of quantification of observations—even a simple count of how many times something happened. Baxter et al. (2001) illustrated this point by reporting the percentage of times they observed a particular behavior out of the total number of observations of that class of behaviors. A significant aspect of Baxter et al.'s study is that they developed a construct in a way that allowed them to measure it, and thereby gave readers a fairly refined sense of the intensity of the phenomenon. The better studies tended to quantify their observations or to embed them within a theoretical framework.

Prior comments notwithstanding, the case studies examined by the committee provided valuable information about variables that program evaluations should include and about the roles that case studies can play in those evaluations. The variables identified by examining case study results arose primarily because people wondered about what it meant for a particular curriculum to be effective. They wondered why particular curricula—each with its own theories of what students need to learn and of how to support students in learning it, and each implemented in settings that posed constraints on how those theories could be actualized in practice—had the effect they did. Therefore, the variables identified by the committee are potential explanatory variables in an evaluation—explaining why a curriculum had the effect it did and helping to answer the question of whether it was effective in fostering students' mathematics learning.

Moreover, the committee believes that if program evaluations systematically included explanatory variables in their study of curriculum effectiveness, the gap between research and evaluation would be largely erased. Thus evaluation studies would become far more valuable to the educational field. Moreover, the inclusion of explanatory variables would give program adopters more precise information about whether the conditions for effectiveness demanded by a particular curriculum coincide with their own local conditions, commitments and resources coincide. Thus evaluation studies would be a valuable resource for stakeholders *and* researchers.

Finally, the committee believes that evaluation studies should include case studies as a matter of design. Few of the case studies examined in this study were planned and executed as part of a larger program evaluation. Instead, faculty and graduate students who were somehow connected to curriculum projects conducted them as research, but independent research was not a component of an overall evaluation. If case studies were included by design in program evaluations or even planned as a systematic set of cases, we would anticipate a greater aggregation of insights into why some programs are effective under certain conditions and not effective under others. Therefore, over time, the creation of principles for designing curricula to achieve results under specific conditions could be established. Many studies would have been strengthened considerably if the investigators had quantified some of their observations, even at the level of simple coding of frequencies of outcome. Descriptions of how the primary constructs were identified and verified also would be helpful.

SYNTHESIS STUDIES

For the purposes of this study, a synthesis study summarizes several evaluation studies across a particular curriculum, discusses the results drawn from these data, and draws conclusions based on the data and the discussion. The evaluations used in synthesis studies may employ their own quantitative analyses, or they may refer to quantitative analyses in the studies they summarize. Summary studies also may refer to qualitative results. Studies used as data for a particular synthesis study might draw conclusion(s) based, inter alia, on standardized tests, items from national and international assessments, college entrance examinations, specially designed assessments, performance of certain students involved in the study using various methods, observations of teachers and classrooms of students, or survey instruments.

In all, the committee found and analyzed 16 synthesis studies of the curricula discussed in this report. Fifteen were NSF supported and one was a UCSMP study. Eleven of 15 appeared in one source (Senk and Thompson, 2002), 10 of which were about different NSF-funded curricula and 1 about UCSMP. The Senk and Thompson book[1] itself is counted among the synthesis studies because it offers synthesis across some or all of these 11 curricula studies in its introductory and concluding chapters. Most of the

---

[1]Senk and Thompson (2002) were funded by the National Science Foundation, ESI-9729228.

studies in Senk and Thompson include a brief statement of the historical background and theoretical basis for the development of the curriculum, the content covered in the curriculum, a discussion of student outcomes, and a discussion of possible explanations for these outcomes (e.g., Carroll and Isaacs, 2002).

Three of the synthesis studies are summaries of those aspects of evaluation that are related to teacher involvement (Romberg, 1997, 2000; Shafer, in press). The remaining study involved political ramifications of a new curriculum (Schoen, Fey, Hirsch, and Coxford, 1999). Each of the 16 synthesis studies is authored by a senior writer of the curriculum materials or by a person closely allied with the curriculum. Therefore, in addition, there is a need for researchers not connected with the curriculum materials to do this type of research.

Examples of Synthesis Studies

Example from Everyday Mathematics

Carroll and Isaacs (2002) summarized each of six quantitative studies measuring student outcomes. These studies, one of which is a longitudinal study, compared outcomes of students using the Everyday Mathematics (EM) curriculum with those who used other curricula. Data were gathered from standardized and specialized tests and survey instruments. The data were drawn mainly from suburban students. References to the original reports were given. Carroll and Isaacs then synthesized data across these studies to conclude:

> Generally, results indicate the following. First, on more traditional topics, such as fact knowledge and paper-and-pencil computation, EM students perform as well as students in more traditional programs. However, EM students use a greater variety of computation solution methods. Students are especially strong on mental computation. Second, on topics that have been underrepresented in the elementary curriculum—geometry, measurement, data, and so on—EM students score substantially higher than do students in more traditional programs. EM students also generally perform better on questions that assess problem solving, reasoning, and communication. Third, although some districts report a decline in computation, especially in the first year or two of implementation, this is usually offset by gains in other areas. Many districts, moreover, report gains in all areas. On tests that are aligned with the National Council of Teachers of Mathematics (NCTM) Standards, such as the Illinois Goal Assessment Program, EM students nearly always show significant improvement over scores before the curriculum was adopted. (pp. 103-104)

Example from the Systemic Initiative for
Montana Mathematics and Science

Lott et al. (2002) offer an example of the historical background and
basis for the development of the curriculum, and the content covered in the
curriculum. They begin with a brief introduction describing the context for
the creation of the SIMMS curriculum as part of the NSF-funded State
Systemic Initiative in Montana. Then they summarize the history of the
curriculum as growing out of a 1989 national survey, "Integrated Mathematics Project," funded by the Exxon Education Foundation. The article
describes the development of the curriculum, the philosophical underpin-
ning, as well as the aims and goals of the various curriculum levels. The
authors discuss assessments that have been conducted in Montana, Cincinnati, and El Paso, and follow-up surveys with certain college students who
had passed three or more full years of SIMMS Integrated Mathematics
(IM). The authors then state the following conclusions:

> Evidence from most facets of the evaluation shows that study with the
> SIMMS IM curriculum does not limit students' abilities on such standard-
> ized tests as the mathematics portion of the Preliminary Scholastic Apti-
> tude Test (PSAT). Teachers of the SIMMS IM curriculum are preparing
> students very well in the areas of problem solving, reasoning, applica-
> tions, communication, and use of technology. Students do at least as well
> overall in collegiate classes, especially the nondevelopmental classes. Stu-
> dents who must take developmental classes in college are at a disadvan-
> tage when compared with students who studied a more traditional curric-
> ulum, though fewer SIMMS IM students appeared in those courses when
> given the option of not taking them.
>
> The collegiate student interviews suggest that the view of collegiate math-
> ematics is not changing as rapidly, specifically in Montana, as the second-
> ary curriculum is changing. . . . The student interviews also suggest that
> teachers at the secondary level need to continue their learning if they are
> to implement reform curricula. Use of technology, an integrated mathe-
> matics curriculum, and new forms of pedagogy provide a basis for needed
> inservice for current teachers at all levels. (pp. 421-422)

Example from Mathematics in Context

Romberg (1997) synthesizes several studies of the impact on teachers of
the Mathematics in Context (MiC) curriculum. Many of these are case
study analyses and are dissertations from Romberg's home institution, the
University of Wisconsin, Madison. In general, these studies trace the impact
of using MiC materials on the practices of fully certified, experienced,
mainly suburban teachers. The MiC materials presented many challenges to

teachers familiar with using traditional instructional practices: Authority for gaining knowledge was transferred from teacher to student. Organizational and management strategies became a problem for some. Views of students and their capabilities were challenged. Romberg concludes:

> This approach to mathematics teaching "represents, on the whole, a substantial departure from teachers' prior experience, established beliefs, and present practice. Indeed, they hold out an image of conditions of learning for children that their teachers have themselves rarely experienced." Such departures from traditional practices were evident in every classroom in these studies. Clearly these departures are nonroutine forms of teaching new to mathematics teachers, and this should lead to new organizational relationship. (p. 377)

Although this synthesis study addresses only one curriculum program, synthesis studies across programs may help to expand the field and shed light on various topics.

Summary

As Senk and Thompson (2002) point out:

> Researchers investigating the effects of curriculum face many issues, including the following: what questions to ask, what type of research design to employ, how to ensure that students using various curricula are comparable at the start of their experience, how to determine the extent to which teachers implement the curriculum, and what measures to use to determine the effects of the curriculum. (p. 17)

The considerable variation in research design and evaluation methods across studies may pose serious challenges to identifying common themes. However, conclusions drawn from such collective evidence can be compelling. The problem with the studies reviewed is that when the syntheses are all written by senior authors of the curricula, the credibility of the results may be challenged. Although these syntheses provided important sources of integrated data on the programs, we found that they tended to lack critical scrutiny and thus may not convince readers that the authors had sought out and included competing interpretations. A common database of variables that all evaluation studies contained could assist researchers when doing synthesis studies and possibly provide additional reader confidence in the findings.

Furthermore, there was a lack of comparison and contrast across programs to discuss how the contrasting and complementary findings around a common research interest might inform each other. Finally, judging by the evidence presented in this report, there is a need to pay much more attention to the adequacy of design of curricular evaluations. The final review chapter by Kilpatrick in Senk and Thompson (2002) provides a more balanced and challenging representation of what is needed to demonstrate curricular effectiveness.

Nonetheless, the committee encourages synthesis evaluations, and funding agencies should consider supporting them, as a means to build on previous knowledge, to provide a summary of existing studies, to enhance understanding of the effectiveness of the various curricula, to build scientific consensus on certain aspects of education research, and to contribute to theory building.

7

Conclusions and Recommendations

Investigating curricular design and implementation is a complex undertaking, and so is reviewing the evaluations of curricula. The committee has conducted its work within a climate of controversy over whether U.S. children are being well served by their mathematical fare. We worked in a period during which proponents of changes to curriculum and pedagogy are struggling to gain acceptance of those changes and being subjected to intense scrutiny as they do so. If these approaches are fundamentally wrongheaded, criticizing them at this time of precariousness is entirely appropriate. If these approaches are potentially worthy, spurious critiques themselves may cause the experimentation to fail. Between these two extremes are a host of other possibilities. The fundamental question of this study was, What is the quality of the evaluations that were designed to judge the effectiveness of these 19 mathematics curricula? An answer to this question should help us learn how to respond to these debates.

Curricular implementation involves the coordination of a variety of factors at differing levels of a system. Evaluations of curricular implementation should acknowledge this complexity, and yet produce reasonably concise, reliable, valid, and cost-efficient evidence of their effectiveness. Education is not simply a bottom-line phenomenon. Thus the effectiveness of curricula depends not only on a simple average or accumulation of effects across test takers, but on a careful assessment of the distribution of effects across grades and topics, across subgroups over time, and across the myriad of unique regional variations of our nation. Implementation, for its part, is not achieved by a blind execution of procedures, but rather by the development of a community of practitioners competently prepared to make appropriate use of materials and exercise judgment in their use. Furthermore, curriculum design is not a rigid scripting of a scope and sequence, but the presentation of sets of tasks and instructional materials linked to relevant standards that can engage students, build on their previous knowledge, and assist them in gaining the mental discipline and proficiency required of knowledgeable citizens and world-class scholars. Effectiveness should consider all these factors, in terms of both potential impact and associated opportunities and risks, and transform them into a judgment concerning a curricular program. In an age of instantaneous recipes and 10-second sound bites, evaluators should provide potential and actual clients with theory-driven, methodologically astute and sound, and practitioner-informed evaluations on which to base curricular decisions.

The committee held fast to a single commitment, namely, that our greatest contribution would be to clarify the proper elements of an array of evaluation studies designed to judge the effectiveness of mathematics curricula, and to clarify what standards of evidence would need to be met to draw conclusions on effectiveness. The committee does not believe any single study determines effectiveness; however, drawing from what could be learned from previously conducted evaluations, we sought to uncover and present practical, sound, and rigorous evaluation designs that could produce the necessary evidence to resolve the debates, bring to the surface variations in values, and propel us toward better serving the nation's youth. We do not claim that the evaluation framework presented herein is a perfect solution to the problems of assessing curricular effectiveness, but rather view it as a way to take stock of our current knowledge base and stimulate the field to modify or refine.

In building the framework, we drew heavily from the National Research Council (NRC) publication entitled *Scientific Research in Education* (NRC, 2002), in which six qualities of scientific research were identified as crucial:

1. Posing significant questions that can be investigated empirically;
2. Linking research to relevant theory;
3. Using methods that permit direct investigation of the question;
4. Providing a coherent and explicit chain of reasoning;
5. Replicating and generalizing across studies; and
6. Disclosing research to encourage professional scrutiny and critique.

Evaluation, as firmly as research, should answer to this set of principles, except that the replication and generalization are more closely subject to the constraints and conditions specified in the program's design. As with scientific research, evaluators need to ensure that when there are competing theories concerning a phenomenon, they work to rule out alternative hypotheses. *Scientific Research in Education* further advocated the value of the use of multiple methodologies to improve one's chances of understanding the complexity in the phenomena under investigation. To complement this work, the committee argued for approaches that draw from multiple methodologies, involve multidisciplinary roots, recognize the importance of ethics and volition, and acknowledge the dependency of the work on building and maintaining mutually respectful relationships with practitioners. Furthermore, evaluation, like research, benefits from the careful accumulation and synthesis of such work. The best that can be asked for and expected is that such experiments in curricular reform be conducted with great care and sensitivity to the values of the constituencies, that they be monitored and reviewed with careful and thorough evaluations, and that the evidence be examined rigorously with periodic review to ensure continuous improvement. As with any scientific enterprise, the specifics of the approaches will evolve with the understanding of the problems.

In addition, there needs to be a commitment by all evaluators and investigators, including the committee, to a generous, thoughtful, and critical consideration of various possible interpretations of the data and a profound intellectual respect for others also undertaking these studies. It does not serve the public well to dismiss the considerable work represented by both the development of the 19 curricula under examination and the efforts of evaluators to document and study their effects. In fact, given the preponderance of studies regarding the curricula supported by the National Science Foundation (NSF), one should acknowledge and value the role of the NSF in requiring the production of many summative evaluations and related research. These were a byproduct of NSF's role in stimulating the development of significant numbers of alternative approaches to curricula in order to meet the need to address the relatively weak performance of American students in mathematics and address the inequities in mathematics learning. Multiple publishers testified that they followed NSF's lead in undertaking their own development efforts. The NSF's activity has been crucial in making this evaluation of evaluations possible and thereby in propelling the nation toward new insights and standards with regard to the conduct of curricular development and accompanying evaluation.

The history of science concerns not only the accumulation of facts and theories, but also the development of method. Developing method combines both a technical prowess as well as theoretical clarifications and negotiated agreement on what terms means and how to gather evidence on issues. To date, there has been too little focus on how to resolve disputes and how to interpret evidence, and too much fractious commentary dismissing others' perspectives on the basis of anecdote and thin doses of empirical data. The committee saw our charge as a means to stabilize method around feasible, valid, and reliable ways to evaluate the quality of evidence on effectiveness.

The committee proceeded in a systematic way to accumulate the array of studies on these curricula, categorizing those studies into four major methods that could shed light on the determination of effectiveness: content analyses, comparative studies, case studies, and synthesis studies. Other studies and submitted reports provided valuable information on the background or the emerging constructs for curricular implementation, but were not sufficiently relevant to our charge. Within these four categories, subcommittees again scrutinized the evaluations and identified the studies that met adequate standards for that methodology. This task required committee members to articulate those standards in the context of mathematics curricula. Each subcommittee compiled their findings, which are based on a careful review of the evaluation studies. Finally, these findings were submitted to the whole committee for review. Then, the committee as a whole drew relationships among those findings, connected those reviews to the framework, and crafted the conclusions and recommendations.

These 19 curricular projects essentially have been experiments. We owe them a careful reading on their effectiveness. Demands for evaluation may be cast as a sign of failure, but we would rather stress that this examination is a sign of the success of these programs to engage a country in a scholarly debate on the question of curricular effectiveness and the essential underlying question, What is most important for our youth to learn in their studies in mathematics? To summarily blame national decline on a set of curricula whose use has a limited market share lacks credibility. At the same time, to find out if a major investment in an approach is successful and worthwhile is a prime example of responsible policy. In experimentation, success and worthiness are two different measures of experimental value. An experiment can fail and yet be worthy. The experiments that probably should not be run are those in which it is either impossible to determine if the experiment has failed or it is ensured at the start, by design, that the experiment will succeed. The contribution of the committee is intended to help us ascertain these distinctive outcomes.

THE QUALITY OF THE EVALUATIONS

The charge to the committee was "to assess the quality of studies about the effectiveness of 13 sets of mathematics curriculum materials developed through NSF support and six sets of commercially generated curriculum materials." Based on our activities, the final product of our work was to present "the criteria and framework for reviewing the evidence, and indicating whether the currently available data are sufficient for evaluating the efficacy of these materials." Finally, if these data were not sufficiently robust, then the committee was also asked to "develop recommendations about the design of a subsequent project that could result in the generation of more reliable and valid data for evaluating these materials."

In response to our charge, the committee finds that:

The corpus of evaluation studies as a whole across the 19 programs studied does not permit one to determine the effectiveness of individual programs with high degree of certainty, due to the restricted number of studies for any particular curriculum, limitations in the array of methods used, and the uneven quality of the studies.

Therefore, according to our charge, we recommend that:

No second phase of this evaluation review should be conducted to determine the effectiveness of any particular program or set of curricular programs dependent on the current database.

The committee emphasizes that we did not directly evaluate the materials. We present no analysis of results aggregated across studies by naming individual curricular programs because we did not consider the magnitude or rigor of the database for individual programs substantial enough to do so. Nevertheless, there are studies that provide compelling data concerning the effectiveness of the program in a particular context. Furthermore, we do report on individual *studies* and their results to highlight issues of approach and methodology. To remain within our primary charge, which was to evaluate the evaluations, we do not summarize results on the individual programs.

The second part of our charge was to present the criteria and framework for reviewing the evidence. To do so, we have developed a set of definitions of key terms which draw on a framework for evaluating the effectiveness of mathematics curricula. Using these definitions and the framework, we were able to undertake a review of the major categories of evaluation studies. We briefly review the definitions and the framework.

FRAMEWORK AND KEY DEFINITIONS

To guide our review of evaluations of mathematics curricula, the committee developed a "Framework for Evaluating Curricular Effectiveness" (see Figure 3-2). This framework emerged from deliberations of the committee following the testimony of experts in the field at two workshops held during 2003, motivated by the need to find common ways to examine different types of evaluations. It permitted the committee to compare evaluations and consider how to identify and distinguish among the variety of methodologies they employed. The committee recommends that individuals or teams charged with curriculum evaluations conduct studies that make use of the following framework:

> Effectiveness of curriculum materials should be determined through evaluation studies that specify the program under investigation in relation to three major components and their interactions:
>
> 1. The program materials and author's design principles;
> 2. The quality, extent, and means of curricular implementation components; and
> 3. The effects on the quality, breadth, type, and distribution of outcomes of student learning over time.
>
> Evaluation studies should further articulate the research design, measurement, and documentation of the above components, and the analysis of results. Secondary components of systemic factors, intervention strategies, and unanticipated influences should also be considered.

The quality of an evaluation depends on how well it connects these components into a chain of reasoning, evidence, and argument to show the effects of curricular use, and to demonstrate their connection to the treatment under investigation. Studies could also include systematic variation to explore which features of curricula are context dependent and which are context independent.

In applying the framework, one needs to distinguish two different aspects of determining curricular effectiveness. First, a single study should demonstrate that it has obtained a level of scientific validity. Then, for a curricular program to be established as effective, a set of scientifically valid studies should be aggregated and synthesized to yield a judgment of effectiveness. We address each of these aspects in turn.

Based on the framework, the committee identified a set of methodological categories of evaluations. For each category, the committee developed a set of methodological expectations for conducting that type of study. This permitted us to define a *scientifically valid study* as follows:

> For a single curricular evaluation to be *scientifically valid*, it should address the components identified in the "Framework for Evaluating Curricular Effectiveness." In addition, it should conform to the methodological expectations of the appropriate category of evaluation as discussed in the report (content analysis, comparative study, or case study). Other designs are possible but would have to address both the theoretical and methodological considerations specified in the framework.

SCIENTIFICALLY ESTABLISHING CURRICULAR EFFECTIVENESS

Defining scientific validity for individual studies is an essential element of understanding curricular effectiveness. However, curricular effectiveness cannot be established by a single scientifically valid study; instead, a body of studies is needed.

Curricular effectiveness is an integrated judgment based on interpretation of a number of scientifically valid evaluations that combine social values, empirical evidence, and theoretical rationales. Similar to assessing test validity (Messick, 1989, 1995) determining effectiveness is a continuing and evolving process. As the body of studies about a curriculum grows larger, findings about its effectiveness can be enhanced or contravened by new findings, new approaches, new research, and changing social conditions

Furthermore, a single methodology, even replications and variations of a study, is inadequate to establish curricular effectiveness, because some types of critical information will be lacking. For example, a content analysis is important because, through expert review of the curriculum content, it provides evidence about such things as the quality of the learning goals or topics that might be missing from a particular curriculum. But content analysis cannot determine whether that curriculum, when actually implemented in classrooms, achieves better outcomes for students. In contrast, a comparative study can provide evidence of improvement in student learning in real classrooms across different curricula. Yet without the kind of complementary evidence provided in a content analysis, nothing will be known about the quality or comprehensiveness of the content in the curriculum that produced higher scores. Furthermore, neither content analyses nor comparative studies typically provide enough detailed information about the quality of the implementation of a particular curriculum. A case study provides deep insight into issues of implementation; however, by itself, it cannot establish representativeness or causality.

Therefore, the committee concluded that:

No single methodology by itself is sufficient to establish a curricular program's effectiveness. The use of multiple methodologies of evaluation strengthens the determination of effectiveness, provided that each is a scientifically valid study.

This conclusion led the committee to propose the following overarching recommendation:

A curricular program's effectiveness should be ascertained through the use of multiple methods of evaluation, each of which should be a scientifically valid study. Periodic synthesis of the results across evaluation studies should also be conducted.

This is a general principle for the conduct of evaluations in recognition that curricular effectiveness is an integrated judgment, continually evolving, and based on scientifically valid evaluations. The committee further recognized that agencies, curriculum developers, and evaluators need a more explicit standard by which to decide whether federally-funded curricula (or curricula from other sources whose adoption and use may be supported by federal monies) can be considered effective enough to adopt. The committee decided to recommend a rigorous standard to which programs should be held to be *scientifically established as effective*. The standard consists of two parts: (1) specification of the array of methodologies required, along with key characteristics, and (2) criteria to determine when the standard has been achieved. The standard relies on the primary methodologies identified in our review, but we acknowledge the possibility of other configurations, provided they draw on the framework and the definition of scientifically valid studies, and include careful review and synthesis of existing evaluations. We view this as an optimal goal to which the field should strive in the attempt to make confident decisions about the effectiveness of any particular curriculum.

The committee recommends that the following standard be used by agencies, curriculum developers, and evaluators:

For a curricular program to be designated *scientifically established as effective*, a collection of scientifically valid evaluation studies addressing its effectiveness should (1) establish that a curricular program and its implementation produce positive and curricularly valid outcomes for students, and (2) convincingly demonstrate that the positive outcomes are due to the curricular intervention. The collection of studies should employ a combination of the following methodologies, and meet the stated criteria:

a. (required) Content analyses by at least two qualified experts (a Ph.D.-level mathematical scientist and a Ph.D.-level mathematics educator), with identified credentials and statements of preference and bias, with due consideration of the systemic fit of the curricula under examination, explicitly addressing the dimensions identified in the content analysis chapter (Chapter 4). The findings from the content analyses should lead to conclusions of overall approval by the content analysts and include explanations by the curriculum authors concerning exceptions they take to the analysts' reports.

b. (required) Comparative studies using experimental or quasi-experimental designs, identifying the comparative curriculum, and addressing the 10 criteria listed in the comparative studies chapter (Chapter 5). Each comparative study should produce findings that the experimental program produces results that meet or exceed those of a comparative program already designated scientifically established as effective, or document significant positive impact on curricularly valid outcomes and indicators of future student success, or exceed the results of a widely used program at a statistically and educationally significant level. Each comparative study should specify the level and type of generalization that can be drawn from it.

c. (highly desirable) One or more case studies to investigate the relationships among the implementation of the curricular program and the program components, as described in the case study chapter (Chapter 6). The case studies should provide documentation that the implementation and outcomes of the program are closely aligned and consistent with the curricular program components and add to the trustworthiness of implementation and the comprehensiveness and validity of the outcome measures.

d. (required) The final report of a program that is scientifically *established as effective* should link the analyses, specify what they convey about the effectiveness of the curriculum, and stipulate the extent to which the program's effectiveness can be generalized, based on the sample populations studied and any other relevant contextual factors and conditions that limit the claims. This report should be made available to the public.

To ensure the independence and impartiality of summative evaluations, which are necessary to scientifically establish a program as effective, the committee makes the following overarching recommendation:

Summative evaluations should be conducted by independent evaluation teams with no membership by authors of the curriculum materials or persons under their supervision.

Consistent with the evaluation standards established by the Joint Committee on Standards for Educational Evaluation, we recognize that in addition to standards for scientific accuracy, evaluation designs need to take into account the needs and resources of stakeholders. This means that evaluation designs should also respond to the demands for utility, feasibility, and propriety. Evaluators must balance the need for scientific rigor with the need for attention to local contextual variations and stakeholders' issues of utility, feasibility, timeliness, and propriety.

RECOMMENDED PRACTICES FOR THE CONDUCT OF EVALUATION STUDIES

In addition to the recommendations above, the committee identified a number of more specific concerns about the evaluation studies it reviewed.

To address these concerns, the committee recommends that individuals or teams charged with conducting curriculum evaluations should strive toward the following recommended practices.

Implementation Components

In relation to implementation, we expressed concerns that across all the studies, there was a disproportionately high representation of students and classrooms from the suburban areas, with weaker representation from urban and rural schools. To address this concern, the committee recommends that:

Evaluations of curricular effectiveness should be conducted with students who represent the intended audience.

In addition, we noted that it is important that judgments of effectiveness be based on clear knowledge and documentation that the program under investigation was adequately and faithfully implemented. To this end, the committee recommends that:

Evaluations should present evidence that provides reliable and valid indicators of the extent, quality, and type of the implementation of the materials. At a minimum, there should be documentation of the extent of coverage of curricular material (what some investigators referred to as "opportunity to learn") and of the extent and type of professional development provided.

The committee recognized the importance of even more specific information and encourages evaluators to seek methods to gather data on additional implementation components. Because of the expense and difficulty of such documentation, we encourage evaluators to at least address these issues through the use of carefully selected case studies. To this end, we recommend that:

Evaluators are advised to provide reports on other implementation factors. These additional factors could include reports on the assignment of students and differential impacts, instructional quality and type, the beliefs and understandings of teachers and students, documentation of formative or embedded assessments, time and resource allocations, and the influence of parents and interest groups.

Outcome Measures

In reviewing the evaluation studies, the committee concluded that across all the studies, there are some major problems with the measurement of student outcomes. These problems make the determination of comparative curricular effectiveness very difficult, and bring potential confusion into the conclusions. These problems include:

- A large and quite varied set of tests for the measurement of achievement, without a sensible and methodologically sound means to compare them;
- Too many tests that rely exclusively on multiple-choice format, limiting the assessment of the cognitive levels of performance and neglecting the long-term development of student knowledge;
- When tests are administered independent of the regular assessment activities, few means to gauge the level of student motivation to perform;
- Lack of clear delineation of whether the measures of prior performance assess different content and skills, prerequisite skills, or the extent to which the current curricular material is already known, or other nonspecific factors of less obvious relevance to curricular effectiveness;
- Reliance on a total test score of mathematics to make judgments, when such tests tend to be less sensitive to curriculum effects than subtest scores focused around very specific content such as fractions;
- For longitudinal studies, lack of methodology to determine if variation in performance by subtopics, across school years, can be validly compared in relation to the psychometrics of the whole test-equating process; and
- Lack of methodology on how to draw conclusions concerning the distribution of results across student groups, including by prior performance levels, to examine not only gains between subgroups or between comparative curricula, but to examine gains within subgroups using a particular curriculum.

The committee could not solve this myriad of problems concerning the outcome measures used to assess curricular effectiveness. However, we did identify two issues that should be clearly distinguished and addressed in relation to all studies. These were labeled as "curricular validity of measures" and "curricular alignment to systemic factors."

To determine effectiveness, outcome measures should be demonstrated to be sensitive to curricular changes. In addition, those measures should comprehensively sample the curricular objectives in the course, measure the content within those objectives validly, and ensure that teaching to the test (rather than the curriculum) is not feasible or likely. The committee used the term "curricular validity of measures" to refer to these requirements.

To address this concern, the committee recommends that:

At a minimum, one of the outcome measures used to determine curricular effectiveness should have demonstrated curricular validity.

Ensuring that curriculum validity of measures is taken into account becomes complex in evaluations involving the comparison of multiple curricula. In such situations the committee decided that each curriculum examined should use, at a minimum, a set of items (which may be a subset of a test) that has curriculum validity of measures. This implies that if a state test is not aligned to a curriculum, it cannot be used to determine curricular effectiveness.

In the context of No Child Left Behind, it should be clear that in order for programs to establish their credentials as effective or as "scientifically based," they will need to show that they have selected outcome measures that demonstrate curricular validity. Accountability without curricular validity is hollow because it is possible to raise scores by teaching to the test, and thus deny students the opportunities to learn the breadth and depth of the entire curriculum. In addition, if measures only sample from the lower levels of the content, particularly at the high school level, the K-12 sector will not have adequate information on students' preparation for advanced study. In our review of the evaluations of the curricula, our deliberations were hampered by the absence of adequately demonstrated curricular validity in outcome measures.

The committee also recognized the importance of the demonstration of evidence of curricular alignment necessary for school decision makers, and that these may be dependent on local contexts. Reports on outcome measures should identify how they connect to the national, state, and local contexts. We labeled consideration of these issues as those of "curricular alignment with systemic factors." To this end, the committee recommends that:

Evaluations should, when possible and relevant, report on a curricular program's alignment to systemic factors, particularly in relation to the local, state, or national mandatory tests or widely used tests having an impact on student opportunities and future activities

Finally, the committee acknowledges the limitations in basing an evaluation of a complex, multifaceted curriculum on a single outcome measure. Thus, the committee recommends that:

Whenever feasible, multiple forms of student outcomes should be used to assess the effectiveness of a curricular program. Measures should consider persistence in course taking, drop-out or failure rates, as well as multiple measures of a variety of the cognitive skills and concepts associated with mathematics learning.

In Chapters 4 through 6, the committee summarizes the results of the review of the subsets of relevant studies. Our focus was on the methodologies of content analyses (Chapter 4), comparative studies (Chapter 5), and case and synthesis studies (Chapter 6). In this chapter, we synthesize our conclusions across these three areas and make recommendations about the most critical issues that need to be addressed to adequately position evaluations to determine curricular effectiveness.

Content Analyses

The committee recognizes that there is little agreement about what should be included in the conduct of content analyses. There were areas of agreement on the part of evaluators across the content analyses, which included the importance of ensuring that the materials were carefully sequenced, comprehensive, and correct. Most authors situated their analyses in the context of an identified set of standards, either at the state level or in reference to Principles and Standards for School Mathematics (NCTM, 2000). Content errors were reported, particularly in first editions, but all participants in the debates showed willingness and commitment to see these identified and fixed quickly and accurately.

In other areas, the committee found distinct differences in preferences and perspectives on content analyses, and was able to find a set of dimensions that seemed to capture those differences. For instance, the committee acknowledged that content analyses conducted a priori are useful and necessary. In addition, we recognized the value of content analysis studies conducted in situ to assess the feasibility of novel approaches prior to formal pilot studies or field testing. On paper, a curriculum may look comprehensive, correct, and orderly, but study of its practical consequences is necessary to ensure its feasibility, its incorporation of adequate levels of challenge and engagement, and its fit with typical or local resources. In order to assist the field in stabilizing this methodology, we outlined dimensions of content analysis and identified some of the key sources of debate.

In relation to content analyses, the committee recommends that:

Content analyses should be recognized as a form of *connoisseurial* assessments, and thus should be conducted by a variety of scholars, including mathematical scientists, mathematics educators, and mathematics teachers and well-qualified individuals, who should identify their qualifications, values concerning mathematical priorities, and potential sources of bias regarding their execution of content analyses.

Furthermore, the committee recommends that:

A content analysis should clearly indicate the extent to which it addresses the following three dimensions:

1. Clarity, comprehensiveness, accuracy, depth of mathematical inquiry and mathematical reasoning, organization, and balance (disciplinary perspectives).
2. Engagement, timeliness and support for diversity, and assessment (learner-oriented perspectives).
3, Pedagogy, resources, and professional development (teacher- and resource-oriented perspective).

Comparative Studies

The committee examined 95 comparative studies. Nationally, there is difference of opinion as to whether anything can be learned from a corpus of studies that collectively exhibit a variety of methodological flaws. The committee took the position that much can be learned through a careful and rigorous examination of the current database, provided those studies meet criteria for studies identified as "at least minimally methodologically adequate." These criteria required that studies:

• Include quantifiably measurable outcomes such as test scores, responses to specified cognitive tasks of mathematical reasoning, performance evaluations, grades, and subsequent course taking; and
• Provide adequate information to judge the comparability of samples.

In addition, a study must have included at least one of the following additional design elements:

• A report of implementation fidelity or professional development activity;
• Results disaggregated by content strands or by performance by student subgroups; and/or
• Multiple outcome measures or precise theoretical analysis of a measured construct, such as number sense, proof, or proportional reasoning.

A set of 63 studies met these criteria and were closely examined for the lessons they could offer on the conduct of future comparative studies of curricular effectiveness. We conducted this review by studying this set in relation to the seven "critical decision points" identified in our framework (Chapter 5). We then examined the pattern of results among these 63 studies by program category (NSF-supported, University of Chicago School Mathematics Project [UCSMP]), and commercially generated) and subjected these results to a process of filtering to see what standards of rigor seemed to influence them most. Finally, we conducted a more thorough review of the studies in relation to what they revealed about analysis by content strand, by equity, and by the interactions among content strand, equity, and grade band. We concluded that comparative studies need to attend most closely to the following three factors:

1. More rigorous design;
2. More precise measures of content-strand outcomes, especially in relation to curricular validity of measures;
3. Careful sampling of representative groups and examination of outcomes by student subgroups.

The committee recommends that comparative study design should attend specifically to at least the following 10 critical decision points and document how they are addressed in individual studies:

1. More pure experimental studies should be conducted, thus ensuring a better balance of experimental and quasi-experimental studies.

2. In quasi-experimental designs, it is necessary to establish comparability by matching samples or making statistical adjustments, using factors such as prior achievement measures, teacher effects, ethnicity, gender, and socioeconomic status (SES). Other factors in need of such consideration are implementation components, as recommended previously.

3. The selection of the correct unit of analysis is critical to the design of comparative studies to establish independence of observations, in relation to tests of significance. Increasingly sophisticated means of conducting studies should be employed, to take into account the level of the educational system in which experimentation occurs.

4. Gathering data on implementation fidelity is essential for evaluators to gauge the adequacy of implementation. Studies could also include nested designs to support analysis of variation by implementation components.

5. Outcome data should include a variety of measures of the highest quality. These measures should vary by question type (open ended, multiple choice), by type of test (international, national and local) and by relation of testing to everyday practice (formative, summative, high stakes), and ensure curricular validity of measures and assess curricular alignment with systemic factors. Tests should also include content strands to aid disaggregation at the level of major content strands.

6. In planning data analyses, careful consideration should be given to the choice of appropriate statistical tests and their interpretation,

including the possible use of sophisticated methods of examining complex data that are becoming readily available such as hierarchical linear modeling.

7. Reports should include clear statements of the limitations to generalization of the study. These should include indications of limitations in populations sampled, sample size, unique population inclusions or exclusions, and levels of use or attrition. Data should also be disaggregated by gender, race/ethnicity, SES, and performance levels to permit readers to see comparative gains across subgroups among between and within studies.

8. Effect sizes should be reported.

9. Careful attention should also be given to the selection of samples of populations for participation. These samples should be representative of the populations to which one wants to generalize the results.

10. The control group should use an identified comparative curriculum or curricula to avoid comparisons to unstructured instruction.

For the purpose of examining the effect of different methodological variations on the results of the evaluations, the committee coded all outcomes of the comparative study, by program type, into positive and statistically significantly stronger than the comparative program, negative and statistically significantly weaker than the comparison program, or showing no significant difference between the control and comparative group (see Table 5-8). We then subjected these results to filter analysis using the seven critical decision points.

Overall, the filtering results suggest that increased rigor seems to lead in general to less strong outcomes, but never reports of completely contrary results. These results also suggest that in recommending design considerations to evaluators, careful attention should be paid to having evaluators include measures of treatment fidelity, considering the impact on all students as well as one particular subgroup; using the correct unit of analysis; and using multiple tests that are also disaggregated by content strand.

The committee recognizes the value of diverse curricular options and finds continuing experimentation in curriculum development to be essential, especially in light of changes in the conduct and use of mathematics and technology. However, it should be accompanied by rigorous efforts to improve our conduct of comparative studies, strengthening the results by learning from previous efforts.

Case Studies

The committee reviewed a set of 45 case studies and selected 32 of these, based on a set of criteria, for intensive review. We saw an important role for case studies, particularly in articulating the mechanisms behind effects. In particular, the committee recommends that:

> Case studies should stipulate clearly what they are cases of, how claims are produced and backed by evidence, and what events are related or left out and why, and should identify explicit underlying mechanisms to explain a rich variety of research evidence.

It is worth noting that case studies often reveal aspects of program components, implementation components, and interactions among these two that behave differently than intended by program designers, and therefore provide essential insights into curricular effectiveness. This is one reason why case studies are a valuable tool in an evaluator's methodological toolkit. The committee emphasizes that a case study should be conducted as rigorously as any other form of study.

Moreover, the committee believes that if program evaluations systematically included explanatory variables in their study of curriculum effectiveness, the gap between research and evaluation would be largely erased. Thus evaluation studies would become far more valuable to the educational field. Moreover, the inclusion of explanatory variables would give program adopters more precise information about whether the conditions for effectiveness demanded by a particular curriculum coincide with their own local conditions, commitments, and resources. Evaluation studies would thus be a valuable resource for stakeholders *and* researchers.

RECOMMENDATIONS TO FEDERAL AND STATE AGENCIES AND PUBLISHERS

Evaluation studies should be undertaken by a variety of scholars with expertise in the following fields: mathematics, mathematics education research, curriculum development, evaluation, statistics, and measurement. These scholars should design and implement the many facets of the evaluation review, working together as a team with regular consultation from stakeholders, including designers, publishers, teachers, administrators, students, and community members. It is preferable that none of these scholars be closely affiliated with the mathematics curriculum materials under review.

The committee recommends that:

> Major efforts should be made by federal agencies to improve the nation's capacity in mathematics curriculum evaluation. Individuals or teams charged with curriculum evaluation should show evidence of understanding the interdisciplinary nature of the task, and involve mathematics educators, mathematicians, measurement specialists, evaluators, and practitioners.

The committee was asked to review the 13 NSF-supported curricula and 6 sets of commercially generated curricula. We note that there was considerable variation in the type and extent of evaluation material provided across these 19 curricula. The database of evaluations for the NSF-supported curricula and for UCSMP greatly exceeded the database for the commercially generated materials in quantity and quality. In establishing a stronger knowledge base for evaluation, it is essential that responsibility for curricular evaluation be shared among three primary bodies: federal agencies developing curricula, publishers, and state and local districts and schools implementing curricula. The committee believes that the typically modest role of districts and schools in evaluation should become more rigorous and significant if we are to require that curricular excellence become the norm in our decentralized system of education. Our review of district-level data was limited by lack of access to such information and minimal means of quality assurance. Furthermore, district and school personnel could benefit from improved data to help determine how and where to focus professional development, respond to local resources and needs, and inform parents of both professional choices and the reasons for those choices. In some instances, an effort to provide assistance in building local capacity to use and interpret evaluation results may be advised. For each of these bodies (publishers, federal, state, and local), the committee made recommendations regarding the conduct of future evaluations. At the federal level, the committee recommends that:

> **Calls for proposals by federal agencies should include more explicit expectations for evaluation of curricular initiatives and increasing sophistication in methodological choices and quality. No federal agency should provide continued funding for major curricular programs that fail to present evaluation data from well-designed, scientifically valid studies.**

Furthermore, the committee recommends that:

> **A federal agency, such as the National Center for Education Statistics, should develop a program for district- and school-level data collection and maintenance on issues of curricular implementation.**

The committee solicited testimony from publishers, who expressed clear willingness to receive guidance on the conduct of evaluation. Some led the way in articulating evaluation methods that could guide comparative review as well as formative evaluations guiding new editions. Those publishers with innovative approaches or unique approaches tended to argue most vigorously for innovative methodologies and were more likely to offer insights into practices needing overhaul, such as methods of adoption that succumb to financial interest more than they respond to reasoned inquiry and sound knowledge bases. Some publishers failed to submit materials for review or any evaluations, and as reported previously, there were many more reviews available for NSF-supported materials. However, some publishers showed weak understanding of the distinctions between market research and scientific research on effectiveness, reporting surveys of teacher preference as methods of curricular evaluation. As a result, the committee recommends that:

Publishers should (1) differentiate market research from scientifically valid evaluation studies and (2) make such evaluation data available to potential clients who use federal funds to purchase curriculum materials.

Districts and schools are the most likely sources of accurate longitudinal data—a critical element in student performance. Local districts and schools should improve their methods of documenting curricular use and linking it to student outcomes. Districts and schools have and should keep more careful records of teachers' professional development activities related to curricula and content learning, and should systematically ensure that all students have fair opportunities to learn, especially under conditions of mobility. Finally, districts and schools can contribute a great deal to the discussion and debate on the impact of accountability systems and their relationship to curricular validity and implementation. To this end, the committee recommends that:

The federal Department of Education, in concert with state education agencies, should undertake an initiative to provide local and district decision makers with training on how to conduct and interpret valid studies of curricular effectiveness.

In addition, the committee recognized that in order to conduct more secure and reliable evaluations, additional basic research is needed in a number of emerging areas pertinent to curricular effectiveness, as discussed in the framework. For example, during the review of content analyses a number of targets of controversy surfaced, including:

- The breadth of topics across years—and extent of integration, multidisciplinary and/or sequential treatment of subfields of numeration, geometry, algebraic reasoning, probability and statistics, and discrete mathematics;
- The relative emphasis of numeration, symbol manipulation skills, and computation and related conceptual development;
- The value, purpose, and use of contextual problems, modeling approaches, and quantitative literacy activities;
- The emphasis on analytic/symbolic, visual, or numeric approaches;

- The reliance on technology as a tool, and the place of manipulatives;
- The importance and effectiveness of student methods and group work;
- The role of practice and item sequencing;
- The role of the teacher in relation to exposition and coaching; and
- The role of different forms of assessment in student learning and achievement.

Thus the committee recommends that:

> The federal government and/or publishers should conduct multi-disciplinary basic empirical research studies on, but not limited to:
> - The interplay among curricular implementation, professional development, and the forms of support and professional interaction among teachers and administrators at the school level;
> - Methods of observing and documenting the type and quality of instruction;
> - Teacher learning from curriculum materials and implementation;
> - The development of outcome measures at the upper level of secondary education and at the elementary level in non-numeration topics that are valid and precise at the topic level;
> - Methods of parent and community education and involvement; and
> - Targets of curricular content controversy such as the appropriate uses of technology; the relative use of analytic, visual, and numeric approaches; or the integration or segregation of the treatment of sub-fields, such as algebra, geometry, statistics, and others.

Although the committee chose not to recommend the proposed second phase of the review of evaluations, it instead proposes that the nation become much more serious and realistic about what is needed to strengthen our knowledge base on curricular effectiveness in mathematics. We have amassed the relevant studies and classified and summarized them. From a subset of these, we have drawn inferences about the conduct of future evaluations. We have proposed a framework for the subsequent conduct of that work, and argued for the need to engage in intensive and inclusive discussions about how to proceed in directions most likely to succeed. We have called for increased attention to program theory and implementation measures in program evaluation, for improvement in the curricular validity of outcomes measures, for improved use of experimental and quasi-experimental research design and coordination of multiple methodologies, and for a coalition of the federal government, districts and schools, and the commercial sector to build capacity to undertake these studies of mathematics curricular effectiveness.

The committee recognizes the complexity and urgency of the challenge, and argues that we should avoid seemingly attractive, but oversimplified, solutions. Although the corpus of evaluations is not sufficient to directly resolve the debates on curricular effectiveness, we believe that in the controversy surrounding mathematics curriculum evaluation, an opportunity exists to forge solutions through negotiation of perspective, to base our arguments on empirical data informed by theoretical clarity, and to build in a critical degree of coherence that is often missing from curricular choice, that is, feedback from careful, valid, and rigorous study. Our intention in presenting this report is to help the nation to take advantage of this opportunity.

References

CHAPTER 1

Adelman, C. (1999). *Answers in the tool box: Academic intensity, attendance patterns, and bachelor's degree attainment.* Washington, DC: U.S. Department of Education, Office of Educational Research and Improvement.

Campbell, P., Jolly, E., Hoey, L., and Perlman, L. (2002). *Upping the numbers: Using research-based decision making to increase diversity in the quantitative disciplines.* A report commissioned by the GE Fund. Newton, MA: Education Development Center. Available: http://www.ge.com/foundation/GEFund_UppingNumbers.pdf [11/5/03].

Cuban, L. (1992). Curriculum stability and change. In P. Jackson (Ed.), *Handbook for research on curriculum: A project of the American Educational Research Association.* New York: Macmillan.

Goodlad, J. (1984). *A place called school.* New York: McGraw-Hill.

Klein, D., Askey, R., Milgram, J., Wu, H., Scharlemann, M., and Tsang, B. (1999). *An open letter to United States Secretary of Education, Richard Riley.* Available: http://www. mathematicallycorrect.com/riley.htm [8/5/03].

Lutzer, D. J. (2003). Mathematics majors 2002. *Notices of the American Mathematical Society, 50*(2), 235-237.

McKnight, C., Crosswhite, J., Dossey, J., Kifer, L., Swafford, J., Travers, K., and Cooney, T. (1987). *The underachieving curriculum: Assessing U.S. school mathematics from an international perspective.* A national report on the Second International Mathematics Study. Champaign, IL: Stipes.

National Council of Teachers of Mathematics. (1989). *Curriculum and evaluation standards for school mathematics.* Reston, VA: Author.

National Research Council. (2002). *Scientific research in education.* Committee on Scientific Principles for Education Research. R. J. Shavelson and L. Towne (Eds.). Center for Education. Division of Behavioral and Social Sciences and Education. Washington, DC: National Academy Press.

National Research Council. (2003). *BIO 2010: Transforming undergraduate education for future research biologists.* Committee on Undergraduate Biology Education to Prepare Research Scientists for the 21st Century. Board on Life Sciences, Division on Earth and Life Studies. Washington, DC: The National Academies Press.

National Science Foundation. (1989). *Materials for middle school mathematics instruction: Program solicitation.* Arlington, VA: Author, Division of Materials Development, Research, and Informal Science Education.

Schmidt, W., McKnight, C., and Raizen, S. (1996). *A splintered vision: An investigation of U.S. science and mathematics education.* U.S. National Research Center for the Third International Mathematics and Science Study (TIMSS) at Michigan State University. Dordrecht, Netherlands: Kluwer Academic.

Takahira, S., Gonzales, P., Frase, M., and Salganik, L. H. (1998). *Pursuing excellence: A study of U.S. twelfth-grade mathematics and science achievement.* Initial Findings from the Third International Mathematics and Science Study. Washington, DC: U.S. Department of Education, National Center for Education Statistics.

CHAPTER 2

Achieve, Inc. (2002). *Foundations for success: Mathematics expectations for the middle grades.* Available: http://www.achieve.org/dstore.nsf/Lookup/Foundations/$file/Foundations.pdf [12/1/03].

Education Market Research. (2001). *Mathematics market, grades K-8: Teaching methods, textbooks/materials used and needed, and market size.* Rockaway Park, NY: Author. Available: http://www.ed-market.com [11/5/03].

Fuson, K. C., Diamond, A., and Fraivillig, J. L. (n.d.). *Implementation of reform norms in Everyday Mathematics classrooms.* (Unpublished manuscript).

National Council of Teachers of Mathematics. (1989). *Curriculum and evaluation standards for school mathematics.* Reston, VA: Author.

National Science Foundation. (1989). *Materials for middle school mathematics instruction: Program solicitation.* Arlington, VA: Author, Division of Materials Development, Research, and Informal Science Education.

National Science Foundation. (1991). *Instructional materials for secondary school mathematics: Program solicitation and guidelines.* Arlington, VA: Author, Directorate for Education and Human Resources.

Simba Information Inc. (2002). *Print publishing for the school market 2001-2002 (yearly report).* Available: http://www.simbanet.com/products/pr_edusr.html#rpt1 [11/5/03].

Simba Information Inc. (2003). *Monthly newsletter.* Available: http://www.simbanet.com/products/pr_edusr.html#nl1 [11/5/03].

The K-12 Mathematics Curriculum Center. (2002). *Curriculum summaries* (6th ed.). Newton, MA: Education Development Center. Available: http://www2.edc.org/mcc/images/currsum6.pdf [11/5/03].

Thompson, D. R., Witonsky, D., Senk, S. L., Usiskin, Z., and Kaeley, G. (2003). *An evaluation of the second edition of UCSMP geometry.* (Unpublished manuscript).

U.S. Department of Education. (1999). *U.S. Department of Education's mathematics and science expert panel exemplary & promising mathematics program.* Available: http://www.enc.org/professional/federalresources/exemplary/promising/ [11/5/03].

U.S. Department of Education. (2001). *Unpublished data from common core of data, 2000–01.* Washington, DC: National Center for Education Statistics. Available: http://nces.ed.gov/Pubs2003/Hispanics/figures.asp?FigureNumber=2_3b [10/1/03].

Weiss, I. R., Banilower, E. R., McMahon, K. C., and Smith, P. S. (2001). *Report on the 2000 national survey of science and mathematics education.* Chapel Hill, NC: Horizon Research. Available: http://2000survey.horizon-research.com/reports/status/complete.pdf [11/5/03].

CHAPTER 3

Agodini, R., and Dynarski, M. (2001). *Are experiments the only option: A look at dropout prevention programs.* Princeton, NJ: Mathematica Policy Research. Available: http://www.mathematica-mpr.com/3rdLevel/propensityscore.htm [9/11/03].

Boruch, R. F. (1997). *Randomized experiments for planning and evaluation: A practical guide.* Thousand Oaks, CA: Sage.

Campbell, D. T. (1969). Reforms as experiments. *American Psychologist, 24*(April), 409-429.

Campbell, D. T., and Stanley, J. C. (1966). *Experimental and quasi-experimental designs for research.* Skokie, IL: Rand McNally.

Caporaso, J. A., and Roos, L. L. (1973). *Quasi-experimental approaches: Testing theory and evaluating policy.* Evanston, IL: Northwestern University Press.

Chen, H. T. (1990). *Theory driven evaluations.* Thousand Oaks, CA: Sage.

Cook, T. D. (in press). *Beyond advocacy: Putting history and research on research into debates about the merits of social experiments.* Thousand Oaks, CA: Sage.

Cook, T. D., and Campbell, D. T. (1979). *Quasi-experimentation: Design and analysis issues for field settings.* Chicago: Rand McNally.

Cordray, D. S., and Fischer, R. L. (1994). Synthesizing evaluation findings. In J. S. Wholey, H. P. Hatry, and K. E. Newcomer (Eds.), *Handbook of practical program evaluation.* San Francisco: Jossey-Bass.

Cronbach, L. J. (1982). *Designing evaluations of educational and social programs.* San Francisco: Jossey-Bass.

Eisner, E. W. (2001). *The educational imagination: On the design and evaluation of school programs* (3rd Ed.). New York: Macmillan.

Fine, M. (1993). [Ap]parent involvement: Reflections on parents, power and urban public schools. *Teachers College Record, 94*(4), 682-729.

Ingersoll, R. M. (2003, November). *The teacher shortage: A case of wrong diagnosis and wrong prescription.* A presentation to the Mathematical Sciences Education Board on November 7.

Lincoln, Y. S., and Guba, E. G. (1986). *Naturalistic inquiry.* Thousand Oaks, CA: Sage.

Lipsey, M. W. (1997). What can you build with thousands of bricks? Musings on the cumulation of knowledge in program evaluation. *Progress and future directions in evaluation: Perspectives on theory, practice, and methods: New directions for evaluation* (Issue 76, 7-24). San Francisco: Jossey-Bass.

Loucks-Horsley, S., Hewson, P., Love, N., and Stiles, K. (1998). *Designing professional development for teachers of science and mathematics.* Thousand Oaks, CA: Corwin Press.

Ma, L. (1999). *Knowing and teaching elementary mathematics: teachers' understanding of fundamental mathematics in China and the United States.* Mahwah, NJ: Lawrence Erlbaum Associates.

McKnight, C., Crosswhite, J., Dossey, J., Kifer, L., Swafford, J., Travers, K., and Cooney, T. (1987). *The underachieving curriculum: Assessing U.S. school mathematics from an international perspective.* A national report on the Second International Mathematics Study. Champaign, IL: Stipes.

National Center for Education Statistics. (1996). *Student learning, teacher quality, and professional development: Theoretical linkages, current measurement, and recommendations for future data collection.* Washington, DC: U.S. Department of Education. Available: http://nces.ed.gov/pubs96/9628.pdf [11/11/03].

National Center for Education Statistics. (2003). *The condition of education 2003.* Washington, DC: U.S. Department of Education. Available: http://nces.ed.gov/pubs2003/2003067.pdf [11/11/03].

National Commission on Teaching and America's Future. (2003). *No dream denied: A pledge to America's children.* Washington, DC: Author. Available: http://www.nctaf.org/dream/report.pdf [11/10/03].

National Research Council. (1992). *Assessing evaluation studies: The case of bilingual education strategies.* Panel to Review Evaluation Studies of Bilingual Education. M. M. Meyer and S. E. Fienberg (Eds.). Committee on National Statistics, Commission on Behavioral and Social Sciences and Education. Washington, DC: National Academy Press.

National Research Council. (1999a). *Designing mathematics or science curriculum programs: A guide for using mathematics and science education standards.* Committee on Science Education K-12 and the Mathematical Sciences Education Board. Center for Science, Mathematics, and Engineering Education. Washington, DC: National Academy Press.

National Research Council. (1999b). *High stakes: Testing for tracking, promotion, and graduation.* Committee on Appropriate Test Use. J. P. Heubert and R. M. Hauser (Eds.). Board on Testing and Assessment, Commission on Behavioral and Social Sciences and Education. Washington, DC: National Academy Press.

National Research Council. (2001a). *Adding it up: Helping children learn mathematics.* Mathematics Learning Study Committee. J. Kilpatrick, J. Swafford, and B. Findell (Eds.). Center for Education, Division of Behavioral and Social Sciences and Education. Washington, DC: National Academy Press.

National Research Council. (2001b). *Investigating the influence of standards: A framework for research in mathematics, science, and technology education.* Committee on Understanding the Influence of Standards in K-12 Science, Mathematics, and Technology Education. I. R. Weiss, M. S. Knapp, K. S. Hollweg, and G. Burrill (Eds.). Center for Education, Division of Behavioral and Social Sciences and Education. Washington, DC: National Academy Press.

National Research Council. (2002). *Scientific research in education.* Committee on Scientific Principles for Education Research. R. J. Shavelson and L. Towne (Eds.). Center for Education, Division of Behavioral and Social Sciences and Education. Washington, DC: The National Academies Press.

National Science Foundation. (1989). *Materials for middle school mathematics instruction: Program solicitation.* Arlington, VA: Author, Division of Materials Development, Research, and Informal Science Education.

Orfield, G., and Kornhaber, M. (2001). *Raising standards or raising barriers: Inequity and high stakes testing in public education.* New York: The Century Foundation.

Porter, A. C. (1995). The uses and misuses of opportunity-to-learn standards. *Educational Researcher, 24*(1), 21-27.

Rossi, P., Freeman, H. E., and Lipsey, M. W. (1999). *Evaluation: A systematic approach.* Thousand Oaks, CA: Sage.

Sconiers, S., Isaacs, A., Higgins, T., McBride, J., and Kelso, C. R. (2002). *Three-state student achievement study project report.* A report by the Arc Center at the Consortium for Mathematics and Its Applications, Boston, MA. (Unpublished manuscript).

Stigler, J. W., and Hiebert, J. (1999). *The teaching gap: Best ideas from the world's teachers for improving education in the classroom.* New York: The Free Press.

Thompson, D. R., Witonsky, D., Senk, S. L., Usiskin, Z., and Kaeley, G. (2003). *An evaluation of the second edition of UCSMP geometry.* (Unpublished manuscript).

Weiss, C. H. (1997). Theory-based evaluation: Past, present, and future. In D. J. Rog and D. Fournier (Eds.), *Progress and future directions in evaluation: Perspectives on theory, practice, and methods: New directions for evaluation* (Issue 76, 40-55). San Francisco: Jossey-Bass.

Yin, R. K. (1994). *Case study research: Design and methods* (2nd edition). Thousand Oaks, CA: Sage.

Yin, R. K. (1997). The abridged version of case study research. In L. Bickman and D. Rog (Eds.), *Handbook of applied social research methods.* Thousand Oaks, CA: Sage.

Yin, R. K., and Bickman, L. (2000). Reforms as non-experiments: A new paradigm. In L. Bickman (Ed.), *Validity and social experimentation: Donald Campbell's legacy.* Thousand Oaks, CA: Sage.

CHAPTER 4

Adams, L., Tung, K.K., Warfield, V.M., Knaub, K., Mudavanhu, B., and Yong, D. (2000, November). *Middle school mathematics comparisons for Singapore mathematics, Connected Mathematics Program, and Mathematics in Context (including comparisons with the NCTM Principles and Standards 2000).* A report to the National Science Foundation, November 2. Seattle, WA: University of Washington. (Unpublished manuscript).

American Association for the Advancement of Science: Project 2061. (1999a). *Algebra textbooks: A standards-based evaluation.* Washington, DC: Author. Available: http://www.project2061.org/research/textbook/hsalg/criteria.htm [7/14/03].

American Association for the Advancement of Science: Project 2061. (1999b). *Middle grades mathematics textbooks: A benchmarks-based evaluation.* Washington, DC: Author. Available: http://www.project2061.org/tools/textbook/matheval/ [7/14/03].

Bishop, W. (1997). *An evaluation of selected mathematics textbooks.* Available: http://mathematicallycorrect.com/bishop4.htm [7/14/03]. (Unpublished manuscript).

Blank, R. (2004, April). *Findings on alignment of enacted curriculum, standards, and assessments: Implications for school improvement strategies under no child left behind.* Presentation at the meeting of the American Educational Research Association, San Diego, CA.

Braams, B. (2003a). *The many ways of arithmetic in UCSMP Everyday Mathematics.* Available: http://www.math.nyu.edu/mfdd/braams/links/em-arith.html [8/27/03]. (Unpublished manuscript).

Braams, B. (2003b). *Spiraling through UCSMP Everyday Mathematics.* Available: http://www.math.nyu.edu/mfdd/braams/links/em-spiral.html [8/27/03]. (Unpublished manuscript).

Bush, W. (1996). *Kentucky middle grades mathematics teacher network: An evaluation of four middle grades mathematics curriculum projects funded by the National Science Foundation (ESI-9253194).* (Unpublished manuscript).

Carroll, W. M. (2001). *A longitudinal study of children in the Everyday Mathematics curriculum.* (Unpublished manuscript).

Clopton, P., McKeown, E., McKeown, M., and Clopton, J. (1998). *Mathematically correct Algebra 1 reviews.* Available: http://mathematicallycorrect.com/algebra.htm [7/14/03]. (Unpublished manuscript).

Clopton, P., McKeown, E., McKeown, M., and Clopton, J. (1999a). *Mathematically correct fifth grade mathematics review.* Available: http://mathematicallycorrect.com/books5.htm [7/14/03]. (Unpublished manuscript).

Clopton, P., McKeown, E., McKeown, M., and Clopton, J. (1999b). *Mathematically correct second grade mathematics review.* Available: http://mathematicallycorrect.com/books2.htm [7/14/03]. (Unpublished manuscript).

Clopton, P., McKeown, E., McKeown, M., and Clopton, J. (1999c). *Mathematically correct seventh grade mathematics review.* Available: http://mathematicallycorrect.com/books7.htm [7/14/03]. (Unpublished manuscript).

Cobb, P., Confrey, J., diSessa, A., Lehrer, R., and Schauble, L. (2003). Design experiment in education research. *Education Researcher, 32*(1), 9-13.

Eisner, E. W. (2001). *The educational imagination: On the design and evaluation of school programs* (3rd ed.). New York: Macmillan.

Gagne, R. M. (1985). *The conditions of learning and theory of instruction* (4th ed.). New York: Holt, Rinehart and Winston.

Kentucky Department of Education. (1996). *Core content for assessment.* Frankfort, KY: Author. Available: http://www.kde.state.ky.us/KDE/Instructional+Resources/Curriculum+Documents+and+Resources/Core+Content+for+Assessment.htm [11/13/03].

Klein, D. (2000). *Weaknesses of everyday mathematics K-3.* Available: http://www.math.nyu.edu/mfdd/braams/nychold/report-klein-em-00.html [8/27/03]. (Unpublished manuscript).

Kulm, G., Morris, K., and Grier, L. (1999). *Middle grades mathematics textbooks: A benchmark-based evaluation.* Washington, DC: American Association for the Advancement of Science: Project 2061. Available: http://www.project2061.org/tools/textbook/matheval/appendx/appendc.htm [11/13/03].

Milgram, R. J. (2003). *An evaluation of CMP.* Available: ers/milgram/report-on-cmp.html" ftp://math.stanford.edu/pub/papers/milgram/report-on-cmp.html [6/4/03]. (Unpublished manuscript).

National Council of Teachers of Mathematics. (1989). *Curriculum and evaluation standards for school mathematics.* Reston, VA: Author.

Porter, A., Flooden, R., Freeman, D., Schmidt, W., and Schwille, J. (1988). Content determinants in elementary school mathematics. In D. A. Grouws and T. J. Cooney (Eds.), *Perspectives on research on effective mathematical teaching* (pp. 96-113). Mahwah, NJ: Lawrence Erlbaum Associates.

Robinson, E., and Robinson, M. (1996). *A guide to standards-based instructional materials in secondary mathematics.* (Unpublished manuscript).

Romberg, T. A., de Lange, J., and Foster, S. (1995). *Welcome to Mathematics in Context: A grade 5 to grade 8 curriculum that meets the NCTM standards.* Madison, WI: University of Wisconsin–Madison.

Schmidt, W., McKnight, C., and Raizen, S. (1996). *A splintered vision: An investigation of U.S. science and mathematics education.* U.S. National Research Center for the Third International Mathematics and Science Study (TIMSS) at Michigan State University. Dordrecht, Netherlands: Kluwer Academic.

Schmidt, W. H., McKnight, C. C., Houang, R. T., Wang, H., Wiley, D., Cogan, L. S., and Wolfe, R. G. (2001). *Why schools matter: A cross-national comparison of curriculum and learning.* San Francisco: Jossey-Bass.

U.S. Department of Education's Mathematics and Science Expert Panel. (1999) *U.S. Department of Education's mathematics and science expert panel exemplary & promising mathematics programs.* Available: http://www.enc.org/professional/federalresources/exemplary/promising/ [7/14/03].

CHAPTER 5

Abrams, B. J. (1989). *A comparison study of the Saxon Algebra I text.* Unpublished doctoral dissertation, University of Colorado at Boulder.

Austin, J., Hirstein, J., and Walen, S. (1997). Integrated mathematics interfaced with science. *School Science and Mathematics, 97*(1), 45-49.

Ben-Chaim, D., Fey, J. T., Fitzgerald, W., Benedetto, C., and Miller, J. (1998). Proportional reasoning among seventh grade students with different curricula experiences. *Educational Studies in Mathematics, 36*(3), 247-273.

Boruch, R. F. (1997). *Randomized experiments for planning and evaluation: A practical guide.* Thousand Oaks, CA: Sage.

Briars, D., and Resnick, L. (2000). *Standards, assessments—and what else? The essential elements of standards-based school improvement.* Los Angeles: UCLA, Center for the Study of Evaluation at the National Center for Research on Evaluation, Standards, and Student Testing. Available: http://www.cse.ucla.edu/CRESST/Reports/TECH528.pdf [8/27/03].

Brown, R. G., Dolciani, M. P., Sorgenfrey, R. H., and Cole, W. L. (1990). *Algebra: Structure and method book–I.* Evanston, IL: McDougal Littel.

Bryk, A. S., and Raudenbush, S. W. (1992). *Hierarchical linear models.* Thousand Oaks, CA: Sage.

Bryk, A. S., Lee, V. E., and Holland, P. B. (1993). *Catholic schools and the common good.* Cambridge, MA: Harvard University Press.

Campbell, D. T., and Stanley, J. C. (1966). *Experimental and quasi-experimental designs for research.* Skokie, IL: Rand McNally.

Caporaso, J. A., and Roos, L. L. (1973). *Quasi-experimental approaches: Testing theory and evaluating policy.* Evanston, IL: Northwestern University Press.

Carroll, W. M. (2001). *A longitudinal study of children in the Everyday Mathematics curriculum.* (Unpublished manuscript).

Carter, A., Beissinger, J., Cirulis, A., Gartzman, M., Kelso, C., and Wagreich, P. (2002). Student learning and achievement with Math Trailblazers. In S. L. Senk and D. R. Thompson (Eds.), *Standards-based school mathematics curricula: What are they? What do students learn?* (pp. 45-78). Mahwah, NJ: Lawrence Erlbaum Associates.

Cohen, J. (1969). *Statistical power analysis for the behavioral sciences.* New York: Academic Press.

Collins, A. M. (2002). *What happens to student learning in mathematics when a multi-faceted, long-term professional development model to support standards-based curricula is implemented in an environment of high stakes testing?* Unpublished doctoral dissertation, Boston College, Boston, MA.

Cook, T. D., and Campbell, D. T. (1979). *Quasi-experimentation: Design and analysis issues for field settings.* Chicago: Rand McNally.

Franke, R. H., and Kaul, J. D. (1978). The Hawthorne experiments: First statistical interpretation. *American Sociological Review, 43*(5), 623-643.

Glass, G. V., McGaw, B., and Smith, M. L. (1981). *Statistical methods for meta-analysis.* New York: Academic Press.

Goodrow, A. (1998). *Children's construction of number sense in traditional, constructivist, and mixed classrooms.* Unpublished doctoral dissertation, Tufts University, Medford, MA.

Hawkins, W. (2003, November). *The Strengthening Underrepresented Minority Mathematics Achievement (SUMMA) program.* Presentation at the meeting of the Mathematical Science Education Board on November 7, Washington, DC.

Heckman, J., and Hotz, J. (1989). Choosing among alternative nonexperimental methods for estimating the impact of social programs. *Journal of the American Statistical Association, 84*(408), 862-880.

Hirsch, C. R., and Schoen, H. L. (2002). *Developing mathematical literacy: A Core-Plus mathematics project longitudinal study progress report.* (Unpublished manuscript).

Huntley, M. A., Rasmussen, C. L., Villarubi, R. S., Sangtong, J., and Fey, J. T. (2000). Effects of standards-based mathematics education: A study of the Core-Plus mathematics project algebra and functions strand. *Journal for Research in Mathematics Education, 31*(3), 328-361.

Joint Committee on Standards for Educational Evaluation. (1994). *The program evaluation standards.* Thousand Oaks, CA: Sage.

Kahan, J. A. (1999). *Relationships among mathematical proof, high school students, and a reform curriculum.* Unpublished doctoral dissertation, University of Maryland at College Park.

Kilpatrick, J. (2002). What works? In S. L. Senk and D. R. Thompson (Eds.), *Standards-oriented school mathematics curricula: What are they? What do students learn?* (pp. 471-488). Mahwah, NJ: Lawrence Erlbaum Associates.

Lipsey, M. W. (1997). What can you build with thousands of bricks? Musings on the cumulation of knowledge in program evaluation. *New Directions for Evaluation, 76,* 7-23.

Lundin, M. A. (2001). *A comparison of former SIMMS and non-SIMMS students on three college-related measures.* Unpublished doctoral dissertation, Montana State University.

Mathison, S., Hedges, L. V., Stodolsky, S., Flores, P., and Sarther, C. (1989). *Teaching and learning algebra: An Evaluation of UCSMP algebra.* (Unpublished manuscript).

McCaffrey, D. F., Hamilton, L. S., Stecher, B. M., Klein, S. P., Bugliari, D., and Robyn, A. (2001). Interactions among instructional practices, curriculum and student achievement: The case of standards-based high school mathematics. *Journal for Research in Mathematics Education, 32*(5), 493-517.

Merlino, F. J., and Wolff, E. (2001). *Assessing the costs/benefits of an NSF "standards-based" secondary mathematics curriculum on student achievement: The Philadelphia experience: Implementing the Interactive Mathematics Program (IMP).* (Unpublished manuscript).

National Council of Teachers of Mathematics. (1989). *Curriculum and evaluation standards for school mathematics.* Reston, VA: Author.

National Research Council. (1992). *Assessing evaluation studies: The case of bilingual education strategies.* Panel to Review Evaluation Studies of Bilingual Education. M. M. Meyer and S. E. Fienberg (Eds.). Committee on National Statistics, Commission on Behavioral and Social Sciences and Education. Washington, DC: National Academy Press.

National Science Foundation. (1988). *Materials for middle school mathematics instruction: Program solicitation.* Arlington, VA: Author, Directorate for Education and Human Resources.

National Science Foundation. (1989). *Materials development, research and informal science education: Program announcement.* Arlington, VA: Author, Division of Materials Development, Research, and Informal Science Education.

National Science Foundation. (1991). *Instructional materials for secondary school mathematics: Program solicitation and guidelines.* Arlington, VA: Author, Directorate for Education and Human Resources.

Peters, K. G. (1992). *Skill performance comparability of two algebra programs on an eighth-grade population.* Unpublished doctoral dissertation, The University of Nebraska–Lincoln.

Peterson, P., Boruch, R., Cook, T., Gueron, J., Hyatt, H., and Mosteller, F. (1999, December). *Can we make education policy on the basis of evidence? What constitutes high quality educational research and how can it be incorporated into policymaking?* Transcript of a Brookings Press Forum, December 8. Washington, DC: Brookings Institution. Available: http://www.brookingsinstitution.org/dybdocroot/comm/transcripts/19991208.htm [8/27/03].

Rentschler, R. V., Jr. (1995). *The effects of Saxon's incremental review of computational skills and problem-solving achievement of sixth-grade students.* Unpublished doctoral dissertation, Walden University.

Riordan, J. E., and Noyce, P. E. (2001). The impact of two standards-based mathematics curricula on student achievement in Massachusetts. *Journal for Research in Mathematical Education, 32*(4), 368-398.

Romberg, T. A., Schafer, M. C., and Webb, N. (in press). The design of the longitudinal / cross-sectional study. In T. A. Romberg and M. C. Schafer (Eds.), *The impact of teaching mathematics using Mathematics In Context on student achievement.*

Rossi, P., Freeman, H. E., and Lipsey, M. W. (1999). *Evaluation: A systematic approach.* Thousand Oaks, CA: Sage.

Saxon, J. (1981). *Algebra I: An incremental development.* Norman, OK: Grassdale.

Schneider, C. (2000). *Connected mathematics and the Texas assessment of academic skills.* Unpublished doctoral dissertation, The University of Texas at Austin.

Schoen, H. L., Hirsch, C. R., and Ziebarth, S. W. (1998, April 15). *An emerging profile of the mathematical achievement of students in the Core-Plus mathematics project.* Paper presented at the Annual Meeting of the American Educational Research Association, San Diego, CA. (Unpublished manuscript).

Schoenfeld, A. (2002). Making mathematics work for all children: Issues of standards, testing, and equity. *Education Researcher, 31*(1), 13-25. Available: http://www.aera.net/pubs/er/pdf/vol31_01/AERA310104.pdf [8/27/03].

Sconiers, S., Isaacs, A., Higgins, T., McBride, J., and Kelso, C. R. (2002). *Three-state student achievement study project report.* A report by the Arc Center at the Consortium for Mathematics and Its Applications, Boston, MA. (Unpublished manuscript).

Secada, W., Fennema, E., and Byrd, L. (1995). *New directions for equity in mathematics education.* New York: Cambridge University Press.

Senk, S. L. (1991, April). *Functions, statistics, and trigonometry with computers at the high school level.* Paper presented at the Annual Meeting of the American Educational Research Association, Chicago, IL. (Unpublished manuscript).

Shafer, M. C. (in press). Expanding classroom practices. In T. A. Romberg (Ed.), *Insight stories: Assessing middle school mathematics.* New York: Teachers College Press.

Souhrada, T. A. (2001). *Secondary school mathematics in transition: A comparative study of mathematics curricula and student results.* Unpublished doctoral dissertation, University of Montana.

Thompson, D. R., Senk, S. L., Witonsky, D., Usiskin, Z., and Kaeley, G. (2001). *An evaluation of the second edition of UCSMP advanced algebra.* (Unpublished manuscript).

Thompson, D. R., Witonsky, D., Senk, S. L., Usiskin, Z., and Kaeley, G. (2003). *An evaluation of the second edition of UCSMP geometry.* (Unpublished manuscript).

U.S. Department of Education. (2001). *No Child Left Behind Act of 2001.* Available: http://www.ed.gov/legislation/ESEA02/107-110.pdf [11/18/03].

Usiskin, Z. (1997). *The evaluation of new curricula.* (Unpublished manuscript).

Walker, R. K. (1999). *Students' conceptions of mathematics and the transition from a standards-based reform curriculum to college mathematics.* Unpublished doctoral dissertation, Western Michigan University.

Wasman, D. (2000). *An investigation of algebraic reasoning of seventh- and eighth-grade students who have studied from the Connected Mathematics curriculum.* Unpublished doctoral dissertation, University of Missouri, Columbia.

Webb, N. L., and Dowling, M. (1995a). *Impact of the Interactive Mathematics Program on the retention of underrepresented students: Class of 1993 transcript report for school 1, Brooks High School.* Project Report 95-3. Madison: University of Wisconsin–Madison, Wisconsin Center for Education Research.

Webb, N. L., and Dowling, M. (1995b). *Impact of the Interactive Mathematics Program on the retention of underrepresented students: Class of 1993 transcript report for school 2, Hill High School.* Project Report 95-4. Madison: University of Wisconsin–Madison, Wisconsin Center for Education Research.

Webb, N. L., and Dowling, M. (1995c). *Impact of the Interactive Mathematics Program on the retention of underrepresented students: Class of 1993 transcript report for school 3, Valley High School.* Project Report 95-5. Madison: University of Wisconsin–Madison, Wisconsin Center for Education Research.

White, P., Gamoran, A., and Smithson, J. (1995). *Math innovations and student achievement in seven high schools in California and New York.* Madison: Consortium for Policy Research (CPRE) and Wisconsin Center for Education Research, School of Education, University of Wisconsin–Madison.

Zahrt, L. T. (2001). *High school reform math programs: An evaluation for leaders.* Unpublished doctoral dissertation, Eastern Michigan University.

CHAPTER 6

Baxter, J., Woodward, J., and Olson, D. (2001). Effects of reform-based mathematics instruction on low-achievers in five third-grade classrooms. *The Elementary School Journal, 101*(5), 529-547.

Bay, J. M. (1999). *Middle school mathematics curriculum implementation: The dynamics of change as teachers introduce and use standards-based curricula.* Unpublished doctoral dissertation, University of Missouri, Columbia.

Bickman, L. (1987). *Using program theory in evaluation.* San Francisco: Jossey-Bass.

Campbell, D. T. (1994). Foreword. In R. K. Yin (Ed.), *Case study research: Design and methods* (2nd ed., pp. ix-xi). Thousand Oaks, CA: Sage.

Carroll, W. M., and Isaacs, A. (2002). Achievement of students using the University of Chicago School Mathematics Project's Everyday Mathematics. In S. L. Senk and D. R. Thompson (Eds.), *Standards-oriented school mathematics curricula: What are they? What do students learn?* (pp. 79-108). Mahwah, NJ: Lawrence Erlbaum Associates.

Cobb, P., Boufi, A., McClain, K., and Whitenack, J. (1997). Reflective discourse and collective reflection. *Journal for Research in Mathematics Education, 28*(3), 258-277.

Cobb, P., Wood, T., Yackel, E., Nicholls, J., Wheatley, G., and Trigatti, B. (1991). Assessment of a problem-centered second grade mathematics project. *Journal for Research in Mathematics Education, 22*(2), 3-29.

Collins, A. M. (2002). *What happens to student learning in mathematics when a multi-faceted, long-term professional development model to support standards-based curricula is implemented in an environment of high stakes testing?* Unpublished doctoral dissertation, Boston College, Boston, MA.

Dapples, B. C. (1994). *Teacher-student interactions in SIMMS and non-SIMMS mathematics classrooms.* Unpublished doctoral dissertation, Montana State University.

de Groot, C. (2000). *Three female voices: The transition to high school mathematics from a reform middle school mathematics program.* Unpublished doctoral dissertation, New York University.

Easley, J. A. Jr. (1977). *On clinical studies in mathematics education.* Washington, DC: U.S. Department of Education. Available: ERIC #: ED146015 [11/20/03].

Fuson, K. C., Diamond, A., and Fraivillig, J. L. (Unknown). *Implementation of reform norms in Everyday Mathematics classrooms.* (Unpublished manuscript).

Greeno, J., and Goldman, S. (1998). *Thinking practices in mathematics and science learning.* Mahwah, NJ: Lawrence Erlbaum Associates.

Herbel-Eisenmann, B. (2000). *How discourse structures norms: A tale of two middle school mathematics classrooms.* Unpublished doctoral dissertation, Michigan State University, East Lansing, MI.

Hetherington, R. A. (2000). *Taking collegial responsibility for implementation of standards-based curriculum: A one-year study of six secondary school teachers.* Unpublished doctoral dissertation, Michigan State University.

Keiser, J., and Lambdin, D. (2001). The clock is ticking: Time constraint issues in mathematics teaching reform. *The Journal of Educational Research, 90*(1), 23-31.

Kett, J. R. (1997). *A portrait of assessment in mathematics reform classrooms.* Unpublished doctoral dissertation, Western Michigan University.

Kramer, S., and Keller, R. (2003). *Tale of synergy: The joint impact of 4 × 4 block scheduling and an NCTM standards-based curriculum on high school mathematics achievement* (DRAFT). (Unpublished manuscript).

Lehrer, R., Jacobson, C., Kemeny, V., and Strom, D. (1999). Building on children's intuitions to develop mathematical understanding of space. In E. Fennema and T. Romberg (Eds.), *Mathematics classrooms that promote understanding* (pp. 63-87). Mahwah, NJ: Lawrence Erlbaum Associates.

Lott, J. W., Hirstein, J., Allinger, G., Walen, S., Burke, M., Lundin, M., Souhrada, T., and Preble, D. (2002). Curriculum and assessment in SIMMS Integrated Mathematics. In S. L. Senk and D. R. Thompson (Eds.), *Standards-oriented school mathematics curricula: What are they? What do students learn?* (pp. 399-423). Mahway, NJ: Lawrence Erlbaum Associates.

Lubienski, S. T. (2000). Problem solving as a means toward mathematics for all: An exploratory look through a class lens. *Journal for Research in Mathematics Education, 31*(4), 454-482.

Manouchehri, A., and Goodman, T. (1998). Mathematics curriculum reform and teachers: Understanding the connections. *The Journal of Educational Research, 92*(1), 27-41.

Manouchehri, A., and Goodman, T. (2000). Implementing mathematics reform: The challenge within. *Educational Studies in Mathematics, 42*, 1-34.

Murphy, L. (1998). *Learning and affective issues among higher- and lower-achieving third-grade students in math reform classrooms: Perspectives of children, parents, and teachers.* Unpublished doctoral dissertation, Northwestern University.

National Research Council. (2002). *Scientific research in education.* Committee on Scientific Principles for Education Research. R. J. Shavelson and L. Towne (Eds.). Center for Education. Division of Behavioral and Social Sciences and Education. Washington, DC: The National Academies Press.

Nicholls, J., Cobb, P., Wood, T., Yackel, E., and Ptashnick, M. (1990). Dimension of success in mathematics: Individual and classroom differences. *Journal for Research in Mathematics Education, 21*, 109-122.

Romberg, T. A. (1997). Mathematics in context: Impact on teachers. In E. Fennema and B. S. Nelson (Eds.), *Mathematics teachers in transition* (pp. 357-380). Mahwah, NJ: Lawrence Erlbaum Associates.

Schoen, H. L., Finn, K. F., Griffin, S. F., and Fi, C. (2003). Teacher variables that relate to student achievement in a standards-oriented curriculum. *Journal for Research in Mathematics Education, 34*(3), 228-259.

Senk, S., and Thompson, D. (2002). *Standards-based school mathematics curricula: What are they? What do students learn?* Mahwah, NJ: Lawrence Erlbaum Associates.

Shafer, M. C. (in press). Expanding classroom practices. In T. A. Romberg (Ed.), *Insight stories: Assessing middle school mathematics.* New York: Teachers College Press.

Smith, S. Z. (1998). *Impact of curriculum reform on a teacher's conception of mathematics.* Unpublished doctoral dissertation, University of Wisconsin, Madison.

Thompson, A. G., Philipp, R. A., Thompson, P. W., and Boyd, B. A. (1994). Calculational and conceptional orientations in teaching mathematics. In A. Coxford (Ed.), *1994 yearbook of the NCTM* (pp. 79-92). Reston, VA: National Council of Teachers of Mathematics.

Woodward, J., and Baxter, J. (1997). The effects of an innovative approach to mathematics on academically low-achieving students in inclusive settings. *Exceptional Children, 63(3)*, 373-388.

Yackel, E., and Cobb, P. (1996). Sociomathematical norms, argumentation, and autonomy in mathematics. *Journal for Research in Mathematics Education, 27*(4), 458-476.

CHAPTER 7

National Council of Teachers of Mathematics. (2000). *Principals and standards for school mathematics*. Reston, VA: Author.

National Research Council. (2002). *Scientific research in education.* Committee on Scientific Principles for Education Research. R. J. Shavelson and L. Towne (Eds.). Center for Education. Division of Behavioral and Social Sciences and Education. Washington, DC: National Academy Press.

Appendix
A

Biographical Sketches

Jere Confrey (Chair) is a professor of mathematics education at Washington University in St. Louis. She is currently vice chair of the Mathematical Sciences Education Board. She was also a member of the National Research Council committee that wrote *Scientific Research in Education.* Dr. Confrey's research has focused on student learning of functions, ratio and proportion, trigonometry, constructivist theory, equity, technology, and urban school reform and systemic change models. She is a co-founder of the UTEACH program for secondary math and science teacher preparation at the University of Texas in Austin, and was the founder of the SummerMath program at Mount Holyoke College and co-founder of SummerMath for Teachers. She is also president of Quest Math and Science Multimedia. She has served as vice president of the International Group for the Psychology of Mathematics Education, as chair of Special Interest Group-Research in Mathematics Education, and as a member of the editorial boards of the *Journal for Research in Mathematics Education, International Journal for Computers in Mathematics Learning,* and *Cognition and Instruction.* Dr. Confrey has taught school at the elementary, secondary, and postsecondary levels. She received a Ph.D. in mathematics education from Cornell University.

Carlos Castillo-Chavez is the Joaquin Bustoz Jr. professor of mathematical biology at Arizona State University. He spent 18 years at Cornell University (1985-2003), the last 4 with a joint professorship appointment in the departments of biological statistics and computational biology and theoretical and applied mechanics. He will continue his association with Cornell University as an adjunct professor. Dr. Castillo-Chavez has received numerous awards, including two White House Awards (1992 and 1997), the 2002 SACNAS Distinguished Scientist Award, and the Richard Tapia Award (2003). He has co-authored more than 100 publications and edited or co-authored six books. His edited volume (with Tom Banks) on the use of mathematical models in homeland security has just been published in Society for Industrial and Applied Mathematics (SIAM) *Frontiers in Applied Mathematics*. He held the position of Ulam Scholar at the Center for Nonlinear Studies at Los Alamos National Laboratory during 2003. He received a Ph.D. in analysis and applied mathematics from the University of Washington, Sevens Point.

Douglas Grouws is a professor of mathematics education at the University of Missouri, where he is a William T. Kemper Fellow. His research focuses on the role of the teacher in facilitating student learning in mathematics in whole-class situations, in small instructional groups, in problem solving, and in the use of technology. He is editor of the *Handbook of Research on Mathematics Teaching and Learning* and has written many articles and chapters on effective mathematics teaching. His current work includes directing the National Science Foundation-funded Mathematics Through Technology Project and serving on a number of committees and boards, including the National Council of Teachers of Mathematics (NCTM) Research Advisory Committee, the Third International Mathematics and Science Study Video Study Advisory Committee, Project Intermath, the NCTM National Assessment of Educational Progress Interpretation Committee, and the *Journal of Mathematics Teacher Education* Editorial Board. He is also a member of the Executive Committee of the Mathematical Sciences Education Board. He received his Ph.D. in mathematics education from the University of Wisconsin.

Carolyn Mahoney is provost and vice chancellor for academic affairs at Elizabeth City State University, North Carolina. She was inducted into the State of Ohio Women's Hall of Fame for Education in 1989. Dr. Mahoney was a member of the steering committee of the Mathematical Education of Teachers Project. She has served on several national programs aimed at improving teacher education in mathematics, as well as enhancing public understanding of and appreciation for mathematics. She is a member of the American Mathematics Society, Mathematical Association of America, National Association of Mathematicians, National Council of Teachers of Mathematics, and Association for Women in Mathematics. Dr. Mahoney earned a Ph.D. in mathematics at Ohio State University in 1983.

Donald Saari is a distinguished professor of mathematics and economics at the University of California, Irvine (UCI) as well as the director of the Institute for Mathematical Behavioral Science and the director of the UCI Center for Decision Analysis. He is a member of the National Academy of Sciences and the previous chair of the U.S. National Committee for Mathematics. He is also a member of the Mathematical Sciences Education Board. In his research Dr. Saari has provided deep analyses of dynamical systems—of Newtonian n-body systems and of classical models of economic equilibrium, showing nonconvergence and modifications that converge—and he has recast voting problems in geometric terms, thereby greatly clarifying the nature of voting paradoxes. He received his Ph.D. in mathematics from Purdue.

William Schmidt is a university distinguished professor and professor of measurement and statistical methods at Michigan State University. He was the national project coordinator and executive director of the center that oversaw U.S. participation in the Third International Mathematics and Science Study. His research interests include statistical modeling and the measurement of curriculum. He has written extensively on comparing U.S. school mathematics curricula with the mathematics curricula of other countries.

Patrick W. Thompson is a professor of mathematics education and the chair of the Department of Teaching and Learning, Peabody College, Vanderbilt University. His research activities are in the areas of algebraic reasoning in elementary and secondary school mathematics development of quantitative reasoning; in mathematics and science the relationships among probabilistic, statistical, and quantitative reasoning; and technology in learning and teaching mathematics. Dr. Thompson has been a member of the Executive Board, International Group for the Psychology of Mathematics Education and co-chair of the Special Interest Group for Research in Mathematics Education of the American Educational Research Association. He has been on numerous editorial boards of key journals in mathematics education and computer science, and has reviewed for journals and government organizations. He received an Ed.D. in mathematics education from the University of Georgia.

William Y. Velez is a distinguished university professor and professor of mathematics at the University of Arizona. His research interests are in number theory and algebraic coding theory. Dr. Velez was given national recognition by receiving the Presidential Award for Excellence in Science, Mathematics and Engineering Mentoring from President Clinton. He is largely responsible for the great number of Hispanic students who have received bachelor's degrees in mathematics from the University of Arizona and for encouraging them to pursue mathematically based careers. He serves on many national advisory committees dealing with the teaching of mathematics at the K-12 level and with minority issues. He received a Ph.D. in mathematics from the University of Arizona.

Appendix
B

Bibliography of Studies Included in Committee Analysis

CONTENT ANALYSIS STUDIES

1. Adams, L., Tung, K. K., Warfield, V. M., Knaub, K., Mudavanhu, B., and Yong, D. (2000). *Middle school mathematics comparisons for Singapore mathematics, Connected Mathematics Program, and Mathematics in Context (including comparisons with the NCTM Principles and Standards 2000)*. Report to the National Science Foundation. Unpublished manuscript.
2. American Association for the Advancement of Science. (1999). *Algebra textbooks: A standards-based evaluation*. Project 2061. Washington, DC: Author.
3. American Association for the Advancement of Science. (1999). *Middle grades mathematics textbooks: A benchmarks-based evaluation*. Project 2061. Washington, DC: Author.
4. Billstein, R. (1998). The STEM model. *Mathematics Teaching in the Middle School, 3*(4), 282-286, 294-296.
5. Bishop, W. (1997). *An evaluation of selected mathematics textbooks*. Available: http://mathematicallycorrect.com/bishop4.htm [7/14/03].
6. Braams, B. (2003). *The many ways of arithmetic in UCSMP Everyday Mathematics*. Available: http://www.math.nyu.edu/mfdd/braams/links/em-arith.html [8/27/03].

7. Braams, B. (2003). *Spiraling through UCSMP everyday mathematics*. Available: http://www.math.nyu.edu/mfdd/braams/links/emspiral.html [8/27/03].

8. Burrill, G., and Romberg, T. A. (1998). Statistics and probability for the middle grades: Examples from mathematics in context. In S. Lajoie (Ed.), *Reflections of statistics: Agendas for learning, teaching, and assessment in K-12*. Mahwah, NJ: Lawrence Erlbaum Associates.

9. Bush, W. (1996). *Kentucky middle grades mathematics teacher network: An evaluation of four middle grades mathematics curriculum projects funded by the National Science Foundation (ESI-9253194)*. Unpublished manuscript.

10. Clopton, P., McKeown, E., McKeown, M. and Clopton, J. (1998). *Mathematically correct algebra 1 reviews*. Available: http://mathematicallycorrect.com/algebra.htm (7/14/03).

11. Clopton, P., McKeown, E., McKeown, M. and Clopton, J. (1999). *Mathematically correct fifth grade mathematics review*. Available: http://mathematicallycorrect.com/books5.htm (7/14/03).

12. Clopton, P., McKeown, E., McKeown, M. and Clopton, J. (1999). *Mathematically correct second grade mathematics review*. Available: http://mathematicallycorrect.com/books2.htm (7/14/03).

13. Clopton, P., McKeown, E., McKeown, M. and Clopton, J. (1999). *Mathematically correct seventh grade mathematics review*. Available: http://mathematicallycorrect.com/books7.htm (7/14/03). Unpublished document.

14. Denny, R. (1993). *STEM evaluation*. Unpublished document.

15. Klein, D. (2000). *Weaknesses of everyday mathematics K-3*. Available: http://www.math.nyu.edu/mfdd/braams/nychold/report-klein-em-00.html [8/27/03]. Unpublished manuscript.

16. Klein, D., and Marple, J. (2000). *A comparison of three K-6 mathematics programs: Sadlier, Saxon, and SRA McGraw-Hill*. Available: http://mathematicallycorrect.com/k6books.pdf [7/14/03].

17. McConnell, J. (1991). *C & D 163 writing assignment program evaluation: UCSMP evaluation Glenbrook South high school*. Unpublished manuscript.

18. McQuire, M., and Simpson, N. (1991). *UCSMP algebra adoption telephone survey, Florida report MR-103-2470*. Unpublished manuscript.

19. McQuire, M., and Simpson, N. (1991). *UCSMP algebra user survey report MR-101-2469*. Unpublished document.

20. Milgram, R. J. (undated). *An evaluation on CMP*. Available: ftp://math.stanford.edu/pub/papers/milgram/report-on-cmp.html [8/27/03].

21. Phillips, E., Lappan, G., Friel, S., and Fey, J. (2001). *Developing coherent high quality curricula: The case of the connected mathematics project.* Working draft of a background paper commissioned for the AAAS Project 2061 Science Textbook Conference, Washington, D.C., February 27-March 2. Unpublished document.

22. Quirk, W. G. (2002). *TERC hands-on math. The truth is in the details: An analysis of investigations in number, data, and space.* Available: http://wgquirk.com/TERC.html.

23. Robinson, E., and Robinson, M. (1996). *A guide to standards-based instructional materials in secondary mathematics.* Unpublished manuscript.

24. Romberg, T., and Pedro, J. D. (1996). *Developing mathematics in context: A research process.* Madison, WI: Wisconsin Center for Education Research.

25. Romberg, T. A., de Lange, J., and Foster, S. (1995). *Welcome to Mathematics in Context: A grade 5 to grade 8 curriculum that meets the NCTM standards.* Madison. University of Wisconsin.

26. Simpson, N. (1991). *Summary of UCSMP Focus Group Meetings.* University of Chicago Users Conference Report MR-103-2484. Unpublished manuscript.

27. Slater, S. (1991-1992). *UCSMP panel final report survey 2 and 3 report MR-103-2515.* Market Research Department, Scott Foresman.

28. Slater, S. (1992). *Teacher lounge simulation, UCSMP teacher's edition report MR-103-2537.* Unpublished manuscript.

29. Slater, S. (1992). *UCSMP panel survey #1 report MR-103-2503.* Market Research Department, Scott Foresman.

30. Slater, S. (1992). *UCSMP panel survey #1, special request data compilation report MR-103-2505.* Market Research Department, Scott Foresman.

31. Slater, S., and Simpson, N. (1992). *UCSMP focus groups report MR-103-2537.* Market Research Department, Scott Foresman.

32. Star, J. R., Herbel-Eisenmann, B. A., and Smith, J. P., III. (2000). Algebraic concepts: What's really new in new curricula? *Mathematics Teaching in the Middle School, 5*(7), 446-451.

33. U.S. Department of Education's Mathematics and Science Expert Panel. (1999). *Exemplary and promising mathematics programs.* Washington, DC: U.S. Department of Education.

34. UCSMP. (1996). *UCSMP user's survey—functions, statistics, and trigonometry.* Chicago. University of Chicago School Mathematics Project.

35. UCSMP. (undated). *UCSMP user's survey—precalculus and discrete mathematics*. Chicago. University of Chicago School Mathematics Project.

36. Wu, H. (2000). *Review of the Interactive Mathematics Program (IMP)*. Available: http://math.berkeley.edu/~wu/IMP2.pdf [8/27/03].

COMPARATIVE STUDIES

1. Abrams, B. J. (1989). *A comparison study of the Saxon algebra I text*. Unpublished doctoral dissertation, University of Colorado at Boulder.

2. Abt Associates, Inc. (Undated). *Independent evaluation of the effectiveness of the math steps curriculum (Houghton Mifflin)*. Unpublished manuscript.

3. Austin Independent School District. (2001). *Austin collaborative for mathematics education, 1999-2000 evaluation*. Unpublished manuscript.

4. Austin, J., Hirstein, J., and Walen, S. (1997). Integrated mathematics interfaced with science. *School Science and Mathematics, 97*(1), 45-49.

5. Bachelis, G. F. (1998). *Reform vs. traditional math curricula: Preliminary report on a survey of the graduating classes of 1997 of Andover high school and Lahser high school, Bloomfield Hills, Michigan, concerning their high school math programs and how well these programs prepared them for college math*. Available: http://www.math.wayne.edu/~greg/original.htm [7/14/03].

6. Ben-Chaim, D., Fey, J. T., Fitzgerald, W., Benedetto, C., and Miller, J. (1998). Proportional reasoning among seventh grade students with different curricula experiences. *Educational Studies in Mathematics, 36*(3), 247-273.

7. Boaler, J. (2002). *Stanford University mathematics teaching and learning study: Initial report: A comparison of IMP 1 and algebra 1 at Greendale School*. Unpublished manuscript.

8. Briars, D., and Resnick, L. (2000). *Standards, assessments—And what else? The essential elements of standards-based school improvement*. Los Angeles, CA: Center for the Study of Evaluation at the National Center for Research on Evaluation, Standards, and Student Testing, UCLA.

9. Calvery, R., Bell, D., and Wheeler, G. (1993, November). *A comparison of selected second and third graders' math achievement: Saxon vs Holt*. Paper presented at the Annual Meeting of the Mid-South Educational Research Association, New Orleans, LA.

10. Carroll, W. M. (1993). *Mathematical knowledge of kindergarten and first-grade students in Everyday Mathematics.* UCSMP Report. Unpublished manuscript.
11. Carroll, W. M. (1994-1995). *Third grade everyday mathematics students' performance on the 1993 and 1994 Illinois state mathematics test.* Unpublished manuscript.
12. Carroll, W. M. (1996). Use of invented algorithms by second graders in a reform mathematics curriculum. *Journal of Mathematical Behavior, 15*(2), 137-150.
13. Carroll, W. M. (1997). Mental and written computation: Abilities of students in a reform-based mathematics curriculum. *The Mathematics Educator, 2*(1), 18-32.
14. Carroll, W. M. (1998). Geometric knowledge of middle school students in a reform-based mathematics curriculum. *School Science and Mathematics, 98*(4), 188-197.
15. Carroll, W. M. (2001). *A longitudinal study of children in the everyday mathematics curriculum.* Unpublished manuscript.
16. Carroll, W. M. (2001). Students in a standards-based curriculum: Performance on the 1999 Illinois state achievement test. *Illinois Mathematics Teacher, 52*(1), 3-7.
17. Carroll, W. M., and Fuson, K. C. (1998). *Multidigit computation skills of second and third graders in everyday mathematics: A follow-up to the longitudinal study.* Unpublished manuscript.
18. Clarke, D., Wallbridge, M., and Fraser, S. (1996). *The other consequences of a problem-based mathematics curriculum.* Unpublished manuscript.
19. Coppola, A. J. (2001). *Evaluation report on SAT scores. MATH Connections: A secondary mathematics core curriculum Southington CT public schools.* Unpublished document.
20. Covington-Clarkson, L. M. (2001). *The effects of the Connected Mathematics Project on middle school mathematics achievement.* Unpublished doctoral dissertation, University of Minnesota, St. Paul.
21. Denson, P. S. (1990). *A comparison of the effectiveness of the Saxon and Dolciani texts and theories about the teaching of high school algebra.* Unpublished doctoral dissertation, Claremont Graduate School.
22. Dowling, M., and Webb, N. L. (1997). *Comparison on a quantitative reasoning test of grade 11 Interactive Mathematics Program (IMP) students with algebra II students at one high school.* Project Report 97-4. University of Wisconsin–Madison. Wisconsin Center for Education Research.
23. Dowling, M., and Webb, N. L. (1997). *Comparison on problem solving and reasoning of grade 10 Interactive Mathematics Program*

(IMP) students with geometry students at one high school. Project Report 97-3. University of Wisconsin–Madison. Wisconsin Center for Education Research.

24. Dowling, M., and Webb, N. L. (1997). *Comparison on statistics items of grade 9 Interactive Mathematics Program (IMP) students with algebra students at one high school.* Project Report 97-2. University of Wisconsin–Madison. Wisconsin Center for Education Research.

25. Drueck, J. V., Fuson, K. C., Carroll, W. M., and Bell, M. S. (1995, April 20-24). *Performance of U.S. first graders in a reform math curriculum compared to Japanese, Chinese, and traditionally taught U.S. students.* Paper presented at the Annual Meeting of the American Education Research Association, San Francisco, CA.

26. Frauenholtz, T. R. (2001). *Relationships among school factors and student mathematics achievement in schools with high and low contact with the SIMMS project.* Unpublished doctoral dissertation, University of Minnesota.

27. Fuson, K. C., and Carroll, W. M. (undated). *Performance of U.S. fifth graders in a reform math curriculum compared to Japanese, Chinese, and traditionally taught U.S. students.* Unpublished manuscript.

28. Fuson, K. C., and Carroll, W. M. (Undated). *Summary of comparison of Everyday Math (EM) and McMillan (MC): Evanston student performance on whole-class tests in grades 1, 2, 3, and 4.* Unpublished manuscript.

29. Fuson, K., Carroll, W., and Drueck, J. (2000). Achievement results for second and third graders using the standards-based curriculum Everyday Mathematics. *Journal for Research in Mathematics Education, 31*(3), 277-295.

30. Glencoe/McGraw-Hill. (Undated). *Study objective and methodology.* New York: Glencoe/McGraw-Hill.

31. Goodrow, A. (1998). *Children's construction of number sense in traditional, constructivist, and mixed classrooms.* Unpublished doctoral dissertation, Tufts University, Medford, MA.

32. Hansen, E., and Greene, K. (2002) *A recipe for math. What's cooking in the classroom: Saxon or traditional?* Available: http://www.secondaryenglish.com/recipeformath.html [8/27/03].

33. Harpster, D. L. (1999). *A study of possible factors that influence the construction of teacher-made problems that assess higher-order thinking skills.* Unpublished doctoral dissertation, Montana State University.

34. Heany, C., Palassis R., and Turner B. (Undated). *A mathematics program for academically gifted sixth graders in district five of Lexington and Richland counties.* Unpublished manuscript.

35. Hill, R., and Parker, T. (2003). *A study of Core-Plus students attending Michigan State University (Draft).* Unpublished manuscript. Available: http://www.math.msu.edu/~hill/HillParker5.pdf [8/27/03]

36. Hirsch, C. R., and Schoen, H. L. (2002). *Developing mathematical literacy: A Core-Plus mathematics project longitudinal study progress report.* Unpublished manuscript.

37. Hirschhorn, D. B. (1991). *Implementation of the first four years of the University of Chicago School Mathematics Project secondary curriculum.* Unpublished doctoral dissertation, University of Chicago.

38. Hirschhorn, D. B., and Senk, S. (1992). Calculators in the UCSMP curriculum for grades 7 and 8. In J. T. Fey and C. R. Hirsch (Eds.), *Calculators in mathematics education.* Reston, VA: National Council of Teachers of Mathematics.

39. Hoover, M. N., Zawojewski, J. S., and Ridgway, J. (1997). *Effects of the Connected Mathematics Project on student attainment.* Unpublished manuscript.

40. Huntley, M. A., Rasmussen, C. L., Villarubi, R. S., Sangtong, J., and Fey, J. T. (2000). Effects of standards-based mathematics education: A study of the Core-Plus mathematics project algebra and functions strand. *Journal for Research in Mathematics Education, 31*(3), 328-361.

41. Johnson, J., Yanyo, L., and Hall, M. (2002). *Evaluation of student math performance in California school districts using Houghton Mifflin mathematics.* Unpublished manuscript.

42. Kahan, J. A. (1999). *Relationships among mathematical proof, high school students, and a reform curriculum.* Unpublished doctoral dissertation, University of Maryland at College Park.

43. Kersaint, G. (1998). *Preservice elementary school teachers' ability to generalize functional relationships.* Unpublished doctoral dissertation, Illinois State University.

44. Lafferty, J. F. (1994). *The links among mathematics text, students' achievement, and students' mathematics anxiety: A comparison of the incremental development and traditional texts.* Unpublished doctoral dissertation, Widener University.

45. Lapan, R., Reys, B., Barnes, D., and Reys, R. (1998). *Standards-based middle grade mathematics curricula: Impact on student achievement.* University of Missouri–Columbia

46. Latterell, C. M. (2000). *Assessing NCTM standards-oriented and traditional students' problem-solving ability using multiple-choice and open-ended questions.* Unpublished doctoral dissertation, University of Iowa.

47. Lawrence, L. K. (1992). *The long-term effects of an incremental development model of instruction upon student achievement and student attitude toward mathematics.* Unpublished doctoral dissertation, University of Tulsa.

48. Leonard, J. D. (1997). *Mathematics reform and the affective domain: Implementing reform at one high school.* Unpublished doctoral dissertation, University of California–Los Angeles.

49. Lundin, M. A. (2001). *A comparison of former SIMMS and non-SIMMS students on three college-related measures.* Unpublished doctoral dissertation, Montana State University.

50. Malouf, S. G. (1999). *A comparison of problem-centered learning model and guided-practice model on high school students' mathematics performance and attitude.* Unpublished doctoral dissertation, University of San Francisco.

51. Mathison, S., Hedges, L. V., Stodolsky, S., Flores, P., and Sarther, C. (1989). *Teaching and learning algebra: An evaluation of UCSMP algebra.* Unpublished manuscript.

52. McCaffrey, D. F., Hamilton, L. S., Stecher, B. M., Klein, S. P., Bugliari, D., and Robyn, A. (2001). Interactions among instructional practices, curriculum and student achievement: The case of standards-based high school mathematics. *Journal for Research in Mathematics Education, 32*(5), 493-517.

53. McConnell, J. (1990). *Performance of UCSMP sophomores on the PSAT Glenbrook South high school.* Unpublished manuscript.

54. Merlino, F. J., and Wolff, E. (2001). *Assessing the costs/benefits of an NSF "standards-based" secondary mathematics curriculum on student achievement: The Philadelphia experience: Implementing the Interactive Mathematics Program (IMP).* Unpublished manuscript.

55. Milgram, R. J. (1999). *Outcomes analysis for Core-Plus students at Andover high school: One year later.* Available: ftp://math.stanford.edu/pub/papers/milgram/andover-report.htm [7/14/03].

56. Milgram, R. J. (1999). *A preliminary analysis of SAT-I mathematics data for IMP schools in California.* Available: http://math.stanford.edu/ftp/milgram/analysis-of-imp-in-california.html [8/27/03].

57. Mokros, J., Berle-Carman, M., Rubin, A., and O'Neil, K. (1996, April 8-12). *Learning operations: Invented strategies that work.* Paper presented at the Annual Meeting of the American Educational Research Association, New York, NY.

58. Mokros, J., Berle-Carman, M., Rubin, A., and Wright, T. (1994). *Full year pilot grades 3 and 4: Investigations in numbers, data, and space.* Cambridge, MA: TERC.

59. Peters, K. G. (1992). *Skill performance comparability of two algebra programs on an eighth-grade population.* Unpublished doctoral dissertation, University of Nebraska-Lincoln.

60. Rentschler, R. V., Jr. (1995). The effects of Saxon's incremental review of computational skills and problem-solving achievement of sixth-grade students. Unpublished doctoral dissertation, Walden University.

61. Reys, R., Reys, B., Lapan, R., Holliday, G., and Wasman, D. (2003). Assessing the impact of standards-based middle grades mathematics textbooks on student achievement. *Journal for Research in Mathematics Education, 34*(1), 74-95.

62. Riordan, J. E., and Noyce, P. E. (2001). The impact of two standards-based mathematics curricula on student achievement in Massachusetts. *Journal for Research in Mathematical Education, 32*(4), 368-398.

63. Riordan, J. E., Noyce, P. E., and Perda, D. (2003, April 21-25). *The impact of two standards-based mathematics curricula on student achievement in Massachusetts: A follow-up study of Connected Mathematics.* Paper Presented at the American Educational Research Association Meeting, Chicago, IL.

64. Roberts, F. H. (1994). *The impact of the Saxon mathematics program on group achievement test scores.* Unpublished doctoral dissertation, The University of Southern Mississippi.

65. Romberg, T. A., Shafer, M. C., and Webb, N. (in press). *The impact of teaching mathematics using Mathematics in Context on student achievement: The design of the longitudinal/cross-sectional study.* Unpublished manuscript.

66. Sanders, B. B. (1999). *The effects of using the Saxon mathematics method of instruction vs. a traditional method of instruction on the achievement of high school juniors.* Georgia Southwestern State University. Available: http://www.gsw.edu/~fspaniol/homepage/7420sanders.PDF [8/27/03].

67. Schneider, C. (2000). *Connected Mathematics and the Texas Assessment of Academic Skills.* Unpublished doctoral dissertation, University of Texas at Austin.

68. Schoen, H. L., and Hirsch, C. R. (2003). Responding to calls for change in high school mathematics: Implications for collegiate mathematics. *The American Mathematical Monthly 110*(2), 109-123.

69. Schoen, H. L., Hirsch, C. R., and Ziebarth, S. W. (1998, April 15). *An emerging profile of the mathematical achievement of students in the Core-Plus mathematics project.* Paper presented at the Annual

Meeting of the American Educational Research Association, San Diego, CA.

70. Schoen, H. L., and Pritchett, J. (1998, April 16). *Students' perceptions and attitudes in a standards-based high school mathematics curriculum.* Paper presented at the Annual Meeting of the American Educational Research Association in San Diego, CA.

71. Sconiers, S., Isaacs, A., Higgins, T., McBride, J., and Kelso, C. R. (2002). *Three-state student achievement study project report (funded by the National Science Foundation).* A Report by The Arc Center at the Consortium for Mathematics and Its Applications (COMAP), Boston, MA. Unpublished manuscript.

72. Segars, J. E. (1994). *Selected factors associated with eighth-grade mathematics achievement.* Unpublished doctoral dissertation, Mississippi State University.

73. Senk, S. L. (1989). Assessing Students' knowledge of functions. In C.A. Mahrer, G.A. Golding, and R. B. Davis (Eds.), *Proceedings of the Eleventh Annual Meeting of the North American Chapter of the International Group for the Psychology of Mathematics Education.*

74. Senk, S. L. (1991). *Functions, statistics, and trigonometry with computers at the high school level.* Paper presented at the Annual Meeting of the American Educational Research Association, Chicago, IL.

75. Sinclair, N. R. W. (1990). *A comparative study of the incremental approach to teaching mathematics and the traditional approach to teaching mathematics.* Unpublished doctoral dissertation, University of Alabama.

76. Sistrunk, K., and Benton, G. (1992, November 11-13). *A comparison of selected fourth graders' math achievement scores after two years in Saxon mathematics: A follow-up study.* Paper presented at the Annual Meeting of the Mid-South Educational Research Association, Knoxville, TN.

77. Souhrada, T. A. (2001). *Secondary school mathematics in transition: A comparative study of mathematics curricula and student results.* Unpublished doctoral dissertation, University of Montana.

78. Staffaroni, M. A. (1996). *Student confidence and perceived usefulness of mathematics: A study of the Math Connections Program.* MATH Connections: A Secondary Mathematics Core Curriculum. Unpublished research paper, Connecticut State Department of Education.

79. Thompson, D. R., and Senk, S. L. (2001). The effects of curriculum on achievement in second-year algebra: The example of the University of Chicago School Mathematics Project. *Journal for Research in Mathematics Education, 32*(1), 58-84.

80. Thompson, D. R., Senk, S. L., Witonsky, D., Usiskin, Z., and Kaeley, G. (2001). *An evaluation of the second edition of UCSMP advanced algebra.* Unpublished manuscript.

81. Thompson, D. R., Witonsky, D., Senk, S. L., Usiskin, Z., and Kaeley, G. (2003). *An evaluation of the second edition of UCSMP geometry.* Unpublished manuscript.

82. Thompson, D. R. (1994). *An evaluation of a new course in precalculus and discrete mathematics.* Unpublished doctoral dissertation, University of Chicago.

83. Tyson, V. V. (1995). *An analysis of the differential performance of girls on standardized multiple-choice mathematics achievement tests compared to constructed response tests of reasoning and problem solving.* Unpublished doctoral dissertation, University of Iowa.

84. Waite, R. D. (2000). *A study of the effects of Everyday Mathematics on student achievement of third-, fourth-, and fifth-grade students in a large north Texas urban school district.* Unpublished doctoral dissertation, University of North Texas.

85. Walker, R. K. (1999). *Students' conceptions of mathematics and the transition from a standards-based reform curriculum to college mathematics.* Unpublished doctoral dissertation, Western Michigan University.

86. Wasman, D. (2000). *An investigation of algebraic reasoning of seventh- and eighth-grade students who have studied form the Connected Mathematics curriculum.* Unpublished doctoral dissertation, University of Missouri, Columbia.

87. Webb, N. L., and Dowling, M. (1995). *Impact of the Interactive Mathematics Program on the retention of underrepresented students: Class of 1993 transcript report for school 1, "Brooks High School."* Project Report 95-3. Madison. University of Wisconsin–Madison. Wisconsin Center for Education Research.

88. Webb, N. L., and Dowling, M. (1995). *Impact of the Interactive Mathematics Program on the retention of underrepresented students: Class of 1993 transcript report for school 2, "Hill High School."* Project Report 95-4. Madison. University of Wisconsin–Madison. Wisconsin Center for Education Research.

89. Webb, N. L., and Dowling, M. (1995). *Impact of the Interactive Mathematics Program on the retention of underrepresented students: Class of 1993 transcript report for school 3, "Valley High School."* Madison, WI: Wisconsin Center for Education Research, University of Wisconsin–Madison.

90. Webb, N. L., and Dowling, M. (1996). *Impact of the Interactive Mathematics Program on the retention of underrepresented students:*

Cross-school analysis of transcripts for the class of 1993 for three high schools. Project Report 96-2. Madison: University of Wisconsin–Madison. Wisconsin Center for Education Research.

91. Webb, N. L., and Dowling, M. (1997). *Replication study of the comparison of IMP students with students enrolled in traditional courses on probability, statistics, problem solving, and reasoning.* Project Report 97-5. Madison: University of Wisconsin–Madison. Wisconsin Center for Education Research.

92. Webb, N. L., and Dowling, M. (1997). *Comparison of IMP students with students enrolled in traditional courses on probability, statistics, problem solving, and reasoning.* Project Report 97-1. Madison, University of Wisconsin–Madison. Wisconsin Center for Education Research.

93. White, P. A., Gamoran, A., and Smithson, J. (1995). *Math innovations and student achievement in seven high schools in California and New York.* University of Wisconsin–Madison. Consortium for Policy Research in Education and the Wisconsin Center for Education Research.

94. Woodward, J., and Baxter, J. (1997). The effects of an innovative approach to mathematics on academically low-achieving students in inclusive settings. *Exceptional Children, 63*(3), 373-388.

95. Zahrt, L. T. (2001). *High school reform math programs: An evaluation for leaders.* Unpublished doctoral dissertation, Eastern Michigan University.

CASE STUDIES

1. Baxter, J., Woodward, J., and Olson, D. (2001). Effects of reform-based mathematics instruction on low-achievers in five third-grade classrooms. *The Elementary School Journal, 101*(5), 529-547.

2. Bay, J., Beem, J., Teys, R., Papick, I., and Barnes, D. (1999). Student reactions to standards-based mathematics curricula: The interplay between curriculum, teachers, and students. *School Science and Mathematics, 99*(4), 182-188.

3. Bay, J. M. (1999). *Middle school mathematics curriculum implementation: The dynamics of change as teachers introduce and use standards-based curricula.* Unpublished doctoral dissertation, University of Missouri, Columbia.

4. Bay, J. M. (2000, April 24-28). *The dynamics of implementing standards-based mathematics curricula in middle schools.* Paper presented at the Annual Meeting of the American Educational Research Association, New Orleans, LA.

5. Carroll, W. M. (1995). *Report on the field test of fifth grade Everyday Mathematics.* UCSMP Report. Unpublished document.

6. Carroll, W. M. (1996). *A follow-up to the fifth-grade field test of Everyday Mathematics: Geometry, and mental and written computation.* UCSMP Report. Unpublished document.

7. Carroll, W. M. (1996). Mental computation of students in a reform-based mathematics curriculum. *School Science and Mathematics, 97*(6), 305-311.

8. Carroll, W. M. (2000). Invented computational procedures of students in a standards-based curriculum. *Journal of Mathematical Behavior, 18*(2), 111-121.

9. Carroll, W. M., and Porter, D. (1994). *A field test of fourth grade Everyday Mathematics.* UCSMP report. Unpublished manuscript.

10. Carter, M. A. (1999). *Student autonomy and making meaning in an urban small school.* Unpublished doctoral dissertation, University of Illinois, Chicago.

11. Collins, A. M. (2002). *What happens to student learning in mathematics when a mutli-faceted, long-term professional development model to support standards-based curricula is implemented in an environment of high stakes testing?* Unpublished doctoral dissertation, Boston College.

12. Dapples, B. C. (1994). *Teacher-student interactions in SIMMS and non-SIMMS mathematics classrooms.* Unpublished doctoral dissertation, Montana State University.

13. De Groot, C. (2000). *Three female voices: The transition to high school mathematics from a reform middle school mathematics program.* Unpublished doctoral dissertation, New York University.

14. Dowling, M. (1996). *Changes in teaching by IMP teachers: A report of findings from a questionnaire administered in 1995.* Unpublished manuscript.

15. Doyle, M. (2000, April 24-28.). *Making meaning of teacher leadership in the implementation of a standards-based mathematics curriculum.* Paper presented at the Annual Meeting of the American Educational Research Association, New Orleans, LA.

16. Drueck, J. V. (1996, April 8-12). *Progression of multidigit addition and subtraction solution methods in high, average, and low math-achieving second graders experiencing a reform curriculum.* Paper presented at the Annual Meeting of the American Education Research Association, New York.

17. Fuson, K. C., Diamond, A., and Fraivillig, J. L. (Undated). *Implementation of reform norms in Everyday Mathematics classrooms.* Unpublished manuscript.

18. Herbel-Eisenmann, B. (2000). *How discourse structures norms: A tale of two middle school mathematics classrooms.* Unpublished doctoral dissertation, Michigan State University, East Lansing.

19. Herbel-Eisenmann, B., Smith, J., and Star, J. (1999, April). *Middle school students' algebra learning: Understanding linear relationships in context.* Discussion draft prepared for the Research Pre-Session of the Annual Meeting of the National Council of Teachers of Mathematics, San Francisco, CA, April 22-24, and the Annual Meeting of the American Educational Research Association, Montreal, Canada.

20. Hetherington, R. A. (2000). *Taking collegial responsibility for implementation of standards-based curriculum: A one-year study of six secondary school teachers.* Unpublished doctoral dissertation, Michigan State University.

21. Hull, L. S. H. (2000). *Teachers' mathematical understanding of proportionality: Links to curriculum, professional development, and support.* Unpublished doctoral dissertation, University of Texas, Austin.

22. Jansen, A., and Herbel-Eisenmann, B. (2001, April 10-14). *Moving from a reform junior high to a traditional high school: Affective, academic, and adaptive mathematical transitions.* Paper presented at the Annual Meeting of the American Educational Research Association, Seattle, WA.

23. Keiser, J., and Lambdin, D. (2001). The clock is ticking: Time constraint issues in mathematics teaching reform. *The Journal of Educational Research, 90*(1), 23-31.

24. Kett, J. R. (1997). *A portrait of assessment in mathematics reform classrooms.* Unpublished doctoral dissertation, Western Michigan University.

25. Kramer, S., and Keller, R. (2003). *Tale of synergy: The joint impact of 4x4 block scheduling and an NCTM standards-based curriculum on high school mathematics achievement.* Unpublished manuscript.

26. Lambdin, D., and Preston, R. (1995). Caricatures in innovation: Teacher adaptation to an investigation-oriented middle school mathematics curriculum. *Journal of Teacher Education, 46*(2), 130-140.

27. Lewis, G., Lazarovici, V., and Smith, J. (2001, April 10-14). *Meeting the demands of calculus and college life: The mathematical experiences of graduates of some reform-based high school programs.* Paper presented at the Annual Meeting of the American Educational Research Association, Seattle, WA.

28. Lubienski, S. T. (1996). *Mathematics for all? Examining issues of class in mathematics teaching and learning.* Unpublished doctoral dissertation, Michigan State University, East Lansing.

29. Lubienski, S. T. (1997, March 24-28). *Successes and struggles of*

striving toward "Mathematics for All": A closer look at socio-economics. Paper presented at the Annual Meeting of the American Educational Research Association, Chicago, IL.

30. Lubienski, S. T. (2000). Problem solving as a means toward mathematics for all: An exploratory look through a class lens. *Journal for Research in Mathematics Education, 31*(4), 454-482.

31. Manouchehri, A., and Goodman, T. (1998). Mathematics curriculum reform and teachers: Understanding the connections. *Journal of Educational Research, 92*(1), 27-41.

32. Manouchehri, A., and Goodman, T. (2000). Implementing mathematics reform: The challenge within. *Educational Studies in Mathematics, 42,* 1-34.

33. Middleton, J. A. (1999). Curricular influences on the motivational beliefs and practice of two middle school mathematics teachers: A follow-up study. *Journal of Research in Mathematics Education, 30*(3), 349-358.

34. Murphy, L. (1998). *Learning and affective issues among higher- and lower-achieving third-grade students in math reform classrooms: Perspectives of children, parents, and teachers.* Unpublished doctoral dissertation, Northwestern University.

35. Pligge, M., Kent, L., and Spence, M. (2000, April 24-28). *Examining teacher change within the context of mathematics curriculum reform: Views from middle school teachers.* Paper presented at the Annual Meeting of the American Educational Research Association, New Orleans, LA.

36. Preston, R. V., and Lambdin, D. V. (1997, March 24-28). *Teachers changing in changing times: Using stages of concern to understand changes resulting from use of an innovative mathematics curriculum.* Paper presented at the Annual Meeting of the American Educational Research Association, Chicago, IL.

37. Schoen, H. L., Finn, K. F., Griffin, S. F., and Fi, C. (2003). Teacher variables that relate to student achievement in a standards-oriented curriculum. *Journal for Research in Mathematics Education, 34*(3), 228-259.

38. Smith, J., and Urdell, B. C. (2001, April 10-14). *"The math is different, but I can deal": Studying students' experiences in a reform-based mathematics curriculum.* Paper presented at the Annual Meeting of the American Educational Research Association, Seattle WA.

39. Smith, J. P., Herbel-Eisenmann, B., Star, J., and Jansen, A. (2000, April 20-21). *Quantitative pathways to understanding using algebra: Possibilities, transitions, and disconnects.* Paper presented at the Research Pre-Session of the National Council of Teachers of Mathematics Annual Meeting, Chicago, IL.

40. Smith, S. Z. (1998). *Impact of curriculum reform on a teacher's conception of mathematics.* Unpublished doctoral dissertation, University of Wisconsin–Madison.

41. Tetley, L., and DuBose, S. (undated). *Problem solving performance of 6th and 7th grade STEM students.* Unpublished master's thesis, University of Missouri.

42. Van Dyke, C. L. (2001). *The shape of things to come: Mathematics reform in the middle school.* Unpublished master's thesis, Pacific Lutheran University.

43. van Reeuwijk, M. (in press). Making instructional decisions: Assessment to inform the teacher. In T. A. Romberg (Ed.), *Insight stories: Assessing middle school mathematics.* New York: Teachers College Press.

44. Webb, D. C. (2000, April 14-28). *Variations in teachers' classroom assessment practices.* Paper presented at the Annual Meeting of the American Educational Research Association, New Orleans, LA.

45. Webb, D. C. (2001). *Instructionally embedded assessment practices of two middle grades mathematics teachers.* Unpublished doctoral dissertation, University of Wisconsin–Madison.

SYNTHESIS STUDIES

1. Billstein, R., and Williamson, J. (2002). Middle grades mathematics: The STEM project. In S. L. Senk and D. R. Thompson (Eds.), *Standards-oriented school mathematics curricula: What are they? What do students learn?* (pp. 251-284). Mahwah, NJ: Lawrence Erlbaum Associates.

2. Carroll, W. M., and Isaacs, A. (2002). Achievement of students using the University of Chicago School Mathematics Project's Everyday Mathematics. In S. L. Senk and D. R. Thompson (Eds.), *Standards-oriented school mathematics curricula: What are they? What do students learn?* (pp. 79-108). Mahwah, NJ: Lawrence Erlbaum Associates.

3. Carter, A., Beissinger, J., Cirulis, A., Gartzman, M., Kelso, C., and Wagreich, P. (2002). Student learning and achievement with Math Trailblazers. In S. L. Senk and D. R. Thompson (Eds.), *Standards-oriented school mathematics curricula: What are they? What do students learn?* (pp. 45-78). Mahwah, NJ: Lawrence Erlbaum Associates.

4. Cichon, D., and Ellis, J. G. (2002). The effects of Math Connections on student achievement, confidence, and perception. In S. L. Senk and D. R. Thompson (Eds.), *Standards-oriented school mathematics*

curricula: What are they? What do students learn? (pp. 345-374). Mahwah, NJ: Lawrence Erlbaum Associates.

5. Lott, J. W., Hirstein, J., Allinger, G., Walen, S., Burke, M., Lundin, M., Souhrada, T., and Preble, D. (2002). Curriculum and assessment in SIMMS Integrated Mathematics. In S. L. Senk and D. R. Thompson (Eds.), *Standards-oriented school mathematics curricula: What are they? What do students learn?* (pp. 399-423). Mahwah, NJ: Lawrence Erlbaum Associates.

6. Mokros, J. (2002). Learning to reason numerically: The impact of Investigations. In S. L. Senk and D. R. Thompson (Eds.), *Standards-oriented school mathematics curricula: What are they? What do students learn?* (pp. 109-131). Mahwah, NJ: Lawrence Erlbaum Associates.

7. Ridgway, J., Zawojewski, J., Hoover, M., and Lambdin, D. (2002). Student attainment in the Connected Mathematics curriculum. In S. L. Senk and D. R. Thompson (Eds.), *Standards-oriented school mathematics curricula: What are they? What do students learn?* (pp. 193-223). Mahwah, NJ: Lawrence Erlbaum Associates.

8. Romberg, T., and Shafer, M. (2002). Mathematics in context: Preliminary evidence about student outcomes. In S. L. Senk and D. R. Thompson (Eds.), *Standards-oriented school mathematics curricula: What are they? What do students learn?* (pp. 225-250). Mahwah, NJ: Lawrence Erlbaum Associates.

9. Romberg, T. A. (1997). Mathematics in context: Impact on teachers. In E. Fennema, and B. S. Nelson (Eds.), *Mathematics teachers in transition* (pp. 357-380). Mahwah, NJ: Lawrence Erlbaum Associates.

10. Romberg, T. A. (2000). *Implementation of Mathematics in Context (MiC): Impact on teachers.* Madison, WI. Unpublished manuscript.

11. Schoen, H., Fey, J. T., Hirsch, C. R., and Coxford, A. F. (1999). Issues and opinions in the math wars. *Phi Delta Kappan, 80*(6), 444-453.

12. Schoen, H. L., and Hirsch, C. R. (2002). *The Core-Plus mathematics project: Perspectives and student achievement.* In S. L. Senk and D. R. Thompson (Eds.), *Standards-oriented school mathematics curricula: What are they? What do students learn?* (pp. 311-343). Mahwah, NJ: Lawrence Erlbaum Associates.

13. Senk, S., and Thompson, D. (2002). *Standards-based school mathematics curricula: What are they? What do students learn?* Mahwah, NJ: Lawrence Erlbaum Associates.

14. Senk, S. L. and. Thompson, D. R. (2003). Effects of the UCSMP secondary curriculum on students' achievement. In S. L. Senk, and D. R. Thompson (Eds.), *Standards-based school mathematics cur-*

ricula: What are they? What do students learn? (pp. 425-456). Mahwah, NJ: Lawrence Erlbaum Associates.

15. Shafer, M. C. (in press). Expanding classroom practices (Chapter 3). In T. A. Romberg (Ed.), *Insight stories: Assessing middle school mathematics.* New York: Teachers College Press.

16. Webb, N. (2002). The impact of the Interactive Mathematics Program on student learning. In S. L. Senk and D. R. Thompson (Eds.), *Standards-oriented school mathematics curricula: What are they? What do students learn?* (pp. 375-398). Mahwah, NJ: Lawrence Erlbaum Associates.

BACKGROUND INFORMATION AND INFORMATIVE STUDIES

Two hundred twenty-five studies were identified as background information or informative studies. These studies were placed in this category because of their potential to shed light on the meaning or interpretation of evaluation data for particular curricula. This category contains the most numerous studies in this review; the distribution of these studies is shown in Table App B-1. They take forms that include dissertations, master's theses and term papers, publisher product promotional materials, unpublished material, and published studies in research or practitioner journals.

Overall, the historical background and informative studies represent more than half of the total studies under review. The committee grouped these studies in the following categories.

- Papers on theories of learning underlying a particular study.
- Data on student or school outcomes or teacher characteristics reported in publishers' descriptions of a particular curriculum. Such data may have been reported by schools using that curriculum and may not have been part of an organized evaluation study.
- Comparative studies that were conducted prior to 1989 because these were listed as background information because much has changed in education since this time (e.g., inception of the National Council of Teachers of Mathematics Standards in1989 and 2000, NSF program solicitations for mathematics instructional materials). These studies provide valuable information, especially in curricula that span the years before and after NSF sponsored the development of mathematics curricula. They offer various philosophies of curriculum design, student achievement data, and potential, and provide insight about how these have changed over time when compared with more current studies of the same curricula.
- Case studies that examine only one curriculum unit and are less than one semester in length.
- Short reports on student achievement in particular districts.

- Use of a particular curriculum to study another concept; for example, the development of students' understanding of angle in a nondirect learning environment.
- Informative studies on teacher responses to a particular curriculum.
- Stories of implementation.
- "How to do it" or curricular implementation discussions by a teacher or school districts.
- Interim or final reports to funding agencies or school districts participating in evaluation studies.
- Book reviews.
- Historical background and program review.
- Curriculum and use of technology.

Although not evaluation studies per se, these studies contribute valuable information about program theory and how decisions were reached that affect curricular design. Reviews of particular curricular programs could find helpful and informative information by reviewing these more closely.

1. Abeille, A., and Hurley, N. (2001). *Final evaluation report, Mathematics: Modeling Our World (MMOW)*. San Francisco: WestEd.
2. Accountability and Development Associates Inc. (1998). *The Arkansas statewide systemic initiative: The ASSI pilot of the Connected Math Project. An evaluation report*. Unpublished manuscript.
3. Alper, L., Fendel, D., Fraser, S., and Resek, D. (1997). Designing a high school curriculum for all students. *American Journal of Education, 106*(1), 148-178.
4. Alper, L., Fendel, D., and Fraser, S. R. D. (1995). Is this a mathematics class? *Mathematics Teacher, 88*(8), 632-638.
5. Anderson, T. (1999, August 2-3). *Using the TI-92 graphing calculator in UCSMP Geometry*. Paper presented at the University of Chicago School Mathematics Project Inservice Conference, Chicago, IL.
6. Arron, D. (1993). *Classroom implementation and impact of Everyday Mathematics K-3: Teachers' perspectives on adopting a reform mathematics curriculum*. Unpublished master's thesis, University of Chicago.
7. Askey, R. (1999). *Knowing and teaching elementary mathematics*. American Educator/American Federation of Teachers. Available: http://www.aft.org/american_educator/fall99/amed1.pdf [7/14/03].
8. Barnard, J. (1995, August). *Sample lessons for UCSMP Algebra Paper*. Presented at the University of Chicago School Mathematics Project Inservice Conference.

9. Ben-Chaim, D., Fey, J., Fitzgerald, W., Benedetto, C., and Miller, J. (1997, March 24-28). *A study of proportional reasoning among seventh and eighth grade students*. Paper presented at the Annual Meeting of the American Education Research Association, Chicago, IL.

10. Billstein, R. (1997). The STEM experience: Some things we've learned and their implication for teacher preparation and inservice. *NCSM Journal of Mathematics Education Leadership, 1*(1), 1-13.

11. Billstein, R. (1998). Middle grades mathematics: The STEM Project—A look at developing a middle school mathematics curriculum. In L. Leutzinger (Ed.), *Mathematics in the middle* (pp. 93-106). Reston, VA: National Council of Teachers of Mathematics.

12. Billstein, R., Williamson, J., et al. (undated). *Six through eight mathematics project design (overview to NSF)*. Unpublished manuscript.

13. Bishop, W. (2003). *Review of standards-based school mathematics curricula: What are they? What do students learn?* Edited by S. L. Senk and D. R. Thompson. Available: http://www.math.nyu.edu/mfdd/braams/nychold/rev-bishop-0302.html [July 15, 2003].

14. Bradfield, P. (1992). *An evaluation of Lamar CISD algebra programs*. Rosenberg, TX: Lamar Consolidated Independent School District.

15. Briars, D. J. (1987). *A comparison of three approaches to algebra I: Applications, incremental and traditional*. Pittsburgh, PA: Pittsburgh Public Schools.

16. Brinker, L. (1996). *Representations and students' rational number reasoning*. Unpublished doctoral dissertation, University of Wisconsin–Madison.

17. Brinker, L. (1998). Using recipes and ratio tables to build on students' understanding of fractions. *Teaching Children Mathematics, 5*(4), 218-224.

18. Brombacher, A. A. (1997). *High school mathematics teachers' transition to a standards-based curriculum*. Unpublished doctoral dissertation, University of Georgia.

19. Budzynski, B. (1994). *Letter of October 5th to Zalman Usiskin regarding scores of students in Ludington (MI) Area Schools on the MEAP tests*. Ludington, MI: Ludington Area Schools.

20. Bussey, J. (2001). Mathematics for the alternative high school student. *Journal of Court, Community, and Alternative Schools* (Spring), pp. 45-51.

21. Calhoun, D. (1996). *Interactive mathematics project progress report 1992-1996*. Fresno, CA: Fresno Unified School District.

22. Carroll, W. M., and Fuson, K. C. (1998). *Computation skills and strategies of second and third graders in Everyday Mathematics:*

Interview results from the longitudinal study. Unpublished manuscript.

23. Celedon, S. (1998). *An analysis of a teacher's and students' language use to negotiate meaning in an ESL/mathematics classroom.* Unpublished doctoral dissertation, University of Texas at Austin.
24. Cichon, D. (1997). *Site visit interim report #3 for Math Connections: Analysis of the year's site visits, 1996-97 school year.* Unpublished manuscript.
25. Clarke, B. A. (1995). *Expecting the unexpected: Critical incidents in the mathematics classroom.* Unpublished doctoral dissertation, University of Wisconsin–Madison.
26. Clarke, D. M. (1993). *Influences on the changing role of the mathematics teacher.* Unpublished doctoral dissertation, University of Wisconsin–Madison.
27. Cole, K., Coffey, J., and Goldman, S. (1999). Using assessment to improve equity in mathematics. *Educational Leadership, 56*(6), 56-58.
28. Connected Mathematics Project. (2001). *Connected Mathematics Project: Research and evaluation summary.* Available: http://www.phschool.com/math/cmp/research_evaluation/ [August 22, 2003].
29. Coxford, A. F., and Hirsch, C. R. (1996). A common core math for all. *Educational Leadership, 53*(8), 22-25.
30. Crawford, J., and Raia, F. (1986). *Analysis of eighth grade math texts and achievement (executive summary and full report).* Unpublished manuscript.
31. Diamond, A., and Fuson, K. C. (1995). *Types of teacher questions in classrooms using a reform mathematics curriculum.* Unpublished manuscript.
32. Doyle, J. (1999). *A review of the ACT mathematics scores.* Sheboygan (WI) Area School District Internal Report. Unpublished manuscript.
33. Everyday Mathematics. (1997). *Mathematics evaluation report, year two.* Chicago: Everyday Learning Corporation/SRA/McGraw-Hill.
34. Everyday Mathematics. (2000). *Everyday Mathematics sourcebook: A guide for parents, teachers, and administrators.* Chicago: Everyday Learning Corporation/SRA/McGraw-Hill.
35. Everyday Mathematics. (2001). *Student performance on the Illinois standards achievement test.* Chicago: Everyday Learning Corporation/SRA/McGraw-Hill.
36. Everyday Mathematics. (2001). *Student performance on the Massachusetts comprehensive assessment system.* Chicago: Everyday Learning Corporation/SRA/McGraw-Hill.

37. Feijs, E. (in press). Constructing a learning environment that pro-
    motes reinvention. In R. Nemirovsky, A. Rosebery, J. Solomon, and
    B. Warren (Eds.), *Everyday matters in science and mathematics:
    Studies of complex classroom events.*

38. Fisher, A. (1998). Fragile future. *Popular Science, 253*(6), 92-98.

39. Flowers, J. (1998). *A study of proportional reasoning as it relates to
    the development of multiplication concepts.* Unpublished doctoral
    dissertation, University of Michigan, Ann Arbor.

40. Fouch, D., and Moore, D. (undated). *Report on advanced placement
    calculus and statistics at Traverse City high schools, MI.* Unpub-
    lished manuscript.

41. Fraivillig, J. (1996). *Case studies and instructional frameworks of
    expert reform mathematics teaching.* Unpublished doctoral disserta-
    tion, Northwestern University.

42. Fraivillig, J., Murphy, L., and Fuson, K. (1999). Advancing children's
    mathematical thinking in Everyday Mathematics classrooms. *Jour-
    nal for Research in Mathematical Education, 30*(2), 148-170.

43. Frykholm, J., and Pittman, M. (2001). Fostering student discourse:
    "Don't ask me! I'm just the teacher!" *Mathematics Teaching in the
    Middle School, 7*(4), 218-221.

44. Garfunkel, S. (2000). *ARISE: Final report for period 8/92–11/98 to
    NSF.* Unpublished manuscript.

45. Glencoe/McGraw-Hill. (2002). *Glencoe algebra 1 learner verifica-
    tion research.* Educational Publishing Research Center.

46. Glencoe/McGraw-Hill. (2002). *Glencoe pre-algebra learner verifica-
    tion research.* Educational Publishing Research Center.

47. Glencoe/McGraw-Hill. (2002). *High school learner verification re-
    search summary.* Author.

48. Goldman, S., Knudsen, J., and Latvala, M. (1998). Engaging middle
    schoolers in and through real-world mathematics. In L. Leutzinger
    (Ed.), *Mathematics in the middle* (pp. 129-140). Reston, VA: Na-
    tional Council of Teachers of Mathematics.

49. Goodrow, A. M. (1998, July). *Modes of teaching and ways of think-
    ing.* Paper presented at the meeting of the International Society for
    the Study of Behavioral Development, Bern, Switzerland.

50. Graue, M. E., and Smith, S. Z. (1993, April 12-16). *Conceptualizing
    assessment from an instructional perspective.* Paper presented at the
    Annual Meeting of the American Educational Research Association,
    Atlanta, GA.

51. Graue, M. E., and Smith, S. Z. (in press). *Shaping assessment through
    instructional innovation.* In T. A. Romberg (Ed.), *Insight stories:
    Assessing middle school mathematics.* New York: Teachers College

Press. Also published in *Journal of Mathematical Behavior, 15*, 113-136, (1996).

52. Greeno, J. G. (1997). The Middle-School Mathematics Through Applications Project group: Theories and practices of thinking and learning to think. *American Journal of Education, 106*(1), 85-127.
53. Griffin, L., Evans, A., Timms, T., and Trowell, J. (2000). *Arkansas grade 8 benchmark exam (1998-99)*. Available: http://www.phschool.com/math/cmp/research_evaluation/data.pdf [8/22/03].
54. Grunow, J. E. (1998). *Using concept maps in a professional development program to assess and enhance teachers' understanding of rational numbers*. Unpublished doctoral dissertation, University of Wisconsin–Madison.
55. Gutstein, E., Lipman, P., Hernandez, P., and de los Reyes, R. (1997). Culturally relevant mathematics teaching in a Mexican American context. *Journal for Research in Mathematics Education, 28*(6), 709-737.
56. Hart, D. (1996). *A tale of two schools: LAUSD and Saxon*. Unpublished manuscript.
57. Hedges, L. V., Stodolsky, S., Flores, P. V., Matheson, D., Sarther, C., and Zhang, J. (1988). *Formative evaluation of UCSMP advanced algebra*. Chicago: University of Chicago School Mathematics Project.
58. Hedges, L. V., Stodolsky, S., Mathison, S., and Flores, P. V. (1986). *Transition mathematics field study*. Chicago: University of Chicago School Mathematics Project.
59. Her, T., and Webb, D. C. (in press). Retracing a path to assessing for understanding. In T. A. Romberg (Ed.), *Insight stories: Assessing middle school mathematics*. New York: Teachers College Press.
60. Hirsch, C. (1998). *Core-Plus Mathematics Project final report to NSF*. Unpublished manuscript.
61. Hirsch, C., Coxford, A., Fey, J., and Schoen, H. (1995). Teaching sensible mathematics in sense-making ways with the CPMP. *Mathematics Teacher, 88*(8), 694-700.
62. Hirsch, C. R., and Coxford, A. F. (1997). Mathematics for all: Perspectives and promising practices. *School Science and Mathematics, 97*(5), 232-241.
63. Holt, Rinehart and Winston. (undated). *Holt middle school math, scientific research base*. Austin, TX: Author.
64. Hull, B. (1999, August 2-3). *UCSMP advanced algebra*. Paper presented at the University of Chicago School Mathematics Project Inservice Conference, Chicago, IL.
65. Hung, C. C. (1995). *Students' reasoning about functions using dependency ideas in the context of an innovative, middle school math-*

ematics curriculum. Unpublished doctoral dissertation, University of Wisconsin–Madison.

66. Hutchinson, E. J. (1998). *Preservice teacher's knowledge: A contrast of beliefs and knowledge of ratio and proportion.* Unpublished doctoral dissertation, University of Wisconsin–Madison.

67. Interactive Mathematics Program. (undated). *Interactive Mathematics Program, phase II: Final summative report to NSF.* Unpublished manuscript.

68. Isaacs, A., Wagreich, P., and Gartzman, M. (1997). The quest for integration: School mathematics and science. *American Journal of Education, 106*(1), 179-206.

69. Isaacs, A. C. W., and. B. M. (2001). *A research-based curriculum: The research basis of the UCSMP Everyday Mathematics curriculum.* Unpublished manuscript.

70. Jakucyn, N. (1999, August 2). *FST for new and experienced teachers.* Paper presented at the University of Chicago School Mathematics Project Inservice Conference, Chicago, IL.

71. Johnson, D. M. and Smith, B. (1981). An evaluation of Saxon's algebra test. *Journal of Educational Research, 81,* 97-102.

72. Kapolka, D. (undated). *Beginners and advanced users of FST and PDM and technology.* Paper presented at the University of Chicago School Mathematics Project Conference.

73. Keiser, J. M. (1997). *The development of students' understanding of angle in a non-directive learning environment.* Unpublished doctoral dissertation, Indiana University, Bloomington.

74. Keiser, J. M. (2000). The role of definition. *Mathematics Teaching in the Middle School, 5*(8), 506-511.

75. Klingele, W. E., and Reed, B. W. (1984). An examination of an incremental approach to mathematics. *Phi Delta Kappan, 65,* 712-713.

76. Koebley, S. C. (1996). *The effects of a constructivist-oriented mathematics classroom on student and parent beliefs about and motivation toward being successful in mathematics.* Unpublished doctoral dissertation, University of Cincinnati.

77. Krebs, A. K. (1999). *Students' algebraic understanding: A study of middle grades students' ability to symbolically generalize functions.* Unpublished doctoral dissertation, Michigan State University, East Lansing.

78. Lambdin, D., and Lappan, G. (1997, April 24-28). *Dilemmas and issues in curriculum reform: Reflections from the Connected Mathematics Project.* Paper presented at the Annual Meeting of the American Education Research Association, Chicago, IL.

79. Lappan, G. (1997). The challenges of implementation: Supporting teachers. *American Journal of Education, 106*(1), 207-239.

80. Lappan, G., and Bouck, M. K. (1998). *Developing algorithms for adding and subtracting fractions. The teaching and learning of algorithms in school mathematics 1998 yearbook.* Reston, VA: National Council of Teachers of Mathematics.

81. Lappan, G., and Phillips, E. (1992). *The first ten months.* Report to NSF. Unpublished manuscript.

82. Lappan, G., and Phillips, E. (1993). *Report of activities for April 1, 1992–April 1, 1993 (ESI-9150217).* Report to NSF. Unpublished manuscript.

83. Lappan, G., and Phillips, E. (1994). *Report of Activities for April 1, 1993–April 1, 1994 (ESI-9150217).* Report to NSF. Unpublished manuscript.

84. Lappan, G., and Phillips, E. (1995). *Report of activities for April 1, 1994–April 1, 1995 (ESI-9150217).* Report to NSF. Unpublished manuscript.

85. Lappan, G., and Phillips, E. (1998). Teaching and learning in the Connected Mathematics Program. In L. Leutzinger (Ed.), *Mathematics in the middle* (pp. 83-92). Reston, VA: National Council of Teachers of Mathematics.

86. Lappan, G., and Phillips, E. (2001). *Connecting teaching, learning and assessment. Final report to NSF May, 2001.* Unpublished manuscript.

87. Leinwand, S. (1996). President's message: "Capturing and sharing success stories." *NCSM Newsletter: Leadership in Mathematics Education, 25*(4), 1-2.

88. Lloyd, G. M., and Wilson, M. (1998). Supporting innovation: The impact of a teacher's conceptions of functions on his implementation of a reform curriculum. *Journal for Research in Mathematics Education, 29*(3), 248-274.

89. Lubienski, S. T. (1997). Class matters: A preliminary excursion. *In Multicultural and gender equity in the mathematics classroom, the gift of diversity 1997 yearbook* (pp. 46-59). Reston, VA: National Council of Teachers of Mathematics.

90. MATH Connections. (undated). *Final and interim reports: Math Connections: A secondary mathematics core curriculum.* Report to NSF. Unpublished manuscript.

91. Math Trailblazers. (undated). *Student achievement: Results, reactions and success stories from the users of Math Trailblazers.* Dubuque, IA: Kendall/Hunt.

92. MathScape Curriculum Center at EDC. (2001). *MathScape: Data from five school systems.* Unpublished manuscript.

93. Mayers, K. S. (1985). *The effects of using the Saxon algebra I textbook on the achievement of ninth-grade algebra students from 1989-1993.* Unpublished doctoral dissertation, Delta State University.

94. McBee, M. (1982). *Dolciani vs. Saxon: A comparison of two Algebra I textbooks with high school students.* Oklahoma City Public Schools. Unpublished manuscript.

95. McDougal Littell. (2002). *The Larson series impact data (Larson, Boswell, Kanold, Stiff).* Houghton Mifflin/McDougal Littell.

96. Meno, L. R. (1995). *Letter to the Texas board of education.* Evaluation report for districts with approved waivers to purchase mathematics textbooks by Saxon Publishers. Unpublished document.

97. Meyer, M. R., Delagardelle, M. L., and Middleton, J. A. (1996). Addressing parents' concerns over curriculum reform. *Educational Leadership, 53*(7), 54-57.

98. Meyer, M. R., and Ludwig, M. A. (1999). Teaching mathematics with MiC: An opportunity for change. *Mathematics Teaching in the Middle School, 4*(4), 264-269.

99. Middleton, J. A. (1993, April 12-16). *The effect of an innovative curriculum project on the motivational beliefs and practice of middle school teachers.* Paper presented at the meeting of the American Educational Research Association, Atlanta, GA.

100. Middleton, J. A. (1994, April 4-8). *Engineering and structural stability as a contextually rich domain for teaching 6th grade geometry.* Paper presented at the Annual Meeting of the American Educational Research Association, New Orleans, LA.

101. Miller, J. (1999). *Report on CMP professional development activities.* Portland, OR. Unpublished manuscript.

102. Nguyen, K., Elam, P., and Weeter, R. K. G. (1993). *The 1992-1993 Saxon mathematics program evaluation report.* Unpublished manuscript.

103. Nguyen, K., and Weeter, R. K. G. (1994). *The 1993-1994 Saxon mathematics program executive summary.* Unpublished manuscript.

104. Phillips, E., Lappan, G., and Grant, Y. (2000). *Implementing standards-based mathematics curricula: Preparing the community, the district, and teachers.* Supported by NSF, ESI-9714999. Unpublished manuscript.

105. Phillips, E. A., Smith, J. P., Star, J., and Herbel-Eisenmann, B. (1998). Algebra in the middle grades. *New England Mathematics Journal, 30*(2), 48-60.

106. Pierce, R. D. (1984). *A quasi-experimental study of Saxon's incremental development model and its effect on student achievement in first-year algebra.* Unpublished doctoral dissertation, University of Tulsa.

107. Platano, D., and Stanziale, L. (1992, November 7-8). *Team Teaching UCSMP to Special Students: A combined effort of an LD certificate and mathematics teacher.* University of Chicago School Mathematics Project Users Conference, Chicago, IL.

108. Plude, M. (1992). *Middlebrook math recommendations (1992-1994) and Transition Math testing results.* Wilton, CT: Middlebrook School.

109. Reed, B. W. (1983). *Incremental, continuous-review versus conventional teaching of algebra.* Unpublished doctoral dissertation, University of Arkansas.

110. Research Communications Limited. (1994). *An evaluation of the STEM sixth grade modules: Executive summary.* Dedham, MA: Author. Unpublished document.

111. Research Communications Limited. (1995). *An evaluation of the STEM seventh grade modules: Summary.* Dedham, MA: Author.

112. Research Communications Limited. (1996). *An evaluation of the STEM eighth grade modules: Summary.* Dedham, MA: Author.

113. Research Communications Limited. (1997). *An evaluation of the STEM curriculum: Sixth, seventh, and eighth grade modules: Summary.* Dedham, MA: Author.

114. Rickard, A. (1993). *Teachers' use of a problem-solving oriented sixth-grade mathematics unit: Two case studies.* Unpublished doctoral dissertation, Michigan State University, East Lansing.

115. Rickard, A. (1995). Teaching with problem-oriented curricula: A case study of middle school mathematics instruction. *Journal of Experimental Education, 64*(1), 5-26.

116. Rickard, A. (1995). Problem solving and computation in school mathematics: Tensions between reforms and practice. *National Forum of Applied Educational Research Journal, 8*(2), 41.

117. Rickard, A. (1996). Connections and confusion: Teaching perimeter and area with a problem-solving oriented unit. *Journal of Mathematical Behavior, 15*(3), 303-327.

118. Rickard, A. (1998). Conceptual and procedural understanding in middle school mathematics. In L. Leutzinger (Ed.), *Mathematics in the middle* (pp. 25-29). Reston, VA: National Council of Teachers of Mathematics.

119. Rodriguez, B. (2000, April 24-28). *An investigation into how a teacher uses a reform-oriented mathematics curriculum.* Paper presented at the Annual Meeting of the American Educational Research Association, New Orleans, LA.

120. Romberg, T. A. (1976). Answering the question "Is it any good?"— The role of evaluation in multi-cultural education through competency-based teacher education. In C. Grant (Ed.), *Sifting and win-*

nowing: An exploration of the relationship between multi-cultural education and CBYTE. Madison, WI: Teacher Corps Associates.

121. Romberg, T. A. (1997). *The development of an "achieved" curriculum for middle school mathematics or Mathematics in Context: A connected curriculum for grades 5-8.* Madison: University of Wisconsin–Madison, National Center for Research in Mathematical Science Education.

122. Romberg, T. A. (1998). *Algebra for the middle grades: An example of cooperative developmental research between American and Dutch scholars.* Paper presented at The Fourth UCSMP International Conference on Mathematics Education, Chicago, IL. Unpublished document.

123. Romberg, T. A. (1998). Designing middle school mathematics materials using problems created to help students progress from informal to formal mathematical reasoning. In L. Leutzinger (Ed.), *Mathematics in the middle* (pp. 107-119). Reston, VA: National Council of Teachers of Mathematics.

124. Romberg, T. A. (1999). Realistic instruction in mathematics. In J. Block, S. Everson, and T. Guskey (Eds.), *Comprehensive school reform* (pp. 287-314). Dubuque, IA: Kendall/Hunt.

125. Romberg, T. A. (2000). *External reviews of Mathematics in Context.* Madison, WI. Unpublished manuscript.

126. Romberg, T. A. (undated). A causal model to monitor changes in school mathematics. In T. A. Romberg and D. M. Stewart (Eds.), *The monitoring of school mathematics: Background papers.* Madison, WI: Wisconsin Center for Education Research.

127. Romberg, T. A., and de Lange, J. (2000). *Realistic mathematics education.* Madison, WI. Unpublished manuscript.

128. Romberg, T. A., and de Lange, J. (in press). Monitoring student progress. In T. A. Romberg and J. de Lange (Eds.), *Insight stories: Assessing middle school mathematics.* New York: Teachers College Press.

129. Romberg, T. A., and Spence, M. S. (1995). Some thoughts on algebra for the evolving work force. In C. Lacampagne, W. Blair, and J. Kaput (Eds.), *The Algebra Initiative Colloquium, Volume 2.* Washington, DC: U.S. Department of Education, Office of Educational Research and Improvement.

130. Romberg, T., and Stewart, D. (1987). *The monitoring of school mathematics: Background papers. Volume 1: The monitoring project and mathematics curriculum.* Unpublished manuscript.

131. Romberg, T. A., Webb, D. C., Burril, J., and Ford, M. J. (2001). *NCISLA middle school design collaborative: Final report to the Verona area school district.* Madison, WI: National Center for Im-

proving Student Learning and Achievement in Mathematics and Science, Wisconsin Center for Education Research.

132. Romberg, T. A., Webb, D. C., Burril, J., and Ford, M. J. (in press). Spreading out the risk for innovation: Building school capacity for teaching for understanding. In T. A. Romberg and T. P. Carpenter (Eds.), *Understanding mathematics and science matters*. Mahwah, NJ: Lawrence Erlbaum Associates.

133. Rush, T. (1996). *A case study of the first year of implementation of the pilot program entitled 'Six Through Eight Mathematics' (STEM)*. Unpublished master's thesis, National-Louis University.

134. Russell, S. J. (2000). *Investigations in Number, Data, and Space: Final report to NSF (ESI-9050210)*. Unpublished manuscript.

135. Sanders, G. (undated). *Letter to Catherine (last name not given) and report regarding results of using Transition Mathematics in Lawrence (KS) school district*. Unpublished document.

136. Saxon, J. (1981). The breakthrough in algebra, II. *National Review*, 1204-1205.

137. Saxon, J. (1981). Incremental development: A breakthrough in mathematics. *Phi Delta Kappan, 63*, 482-484.

138. Saxon Publishers. (2001). *Mathematics results*. Norman, OK: Author.

139. Saxon Publishers. (2002). *Research support: Saxon math*. Norman, OK: Author.

140. Saxon Publishers. (2002). *The 2002 Saxon report card*. Norman, OK: Author.

141. Saxon Publishers. (undated). *Mathematics and phonics: Saxon results*. Norman, OK: Author.

142. Schoen, H. (1993). *Impact study of mathematics education projects funded by the National Science Foundation, 1983-1991: Interactive Mathematics Project report (draft)*. Unpublished manuscript.

143. Schoen, H. (1997). *Core-Plus mathematics project phase II longitudinal study: A brief status report*. Unpublished manuscript.

144. Schoen, H. L., and Ziebarth, S. W. (1997). A progress report on student achievement in the Core-Plus Mathematics Project field test. *NCSM Journal of Mathematics Education Leadership, 1*(3), 15-23.

145. Schoenfeld, A. (2002). Making mathematics work for all children: Issues of standards, testing, and equity. *Education Researcher, 31*(1), 13-25.

146. Senk, S. (1985). How well do students write geometry proofs? *Mathematics Teacher, 78*, 448-456.

147. Senk, S. (1989). Van Hiele levels and achievement in writing geometry proofs. *Journal for Research in Mathematics Education, 20*(3), 309-321.

148. Senk, S. L. (1983). *Proof-writing achievement and Van Hiele levels among secondary school geometry students.* Unpublished doctoral dissertation, University of Chicago.

149. Shafer, M., and Sherian, F. (1997). Changing face of assessment. *Principled Practice in Mathematics and Science Education, 1*(2), 1-8.

150. Shafer, M. C. (1996). *Assessment of student growth in a mathematical domain over time.* Unpublished doctoral dissertation, University of Wisconsin–Madison.

151. Shafer, M. C., and Romberg, T. A. (1999). Assessment in classrooms that promote understanding. In E. Fennema and T. A. Romberg (Eds.), *Mathematics classrooms that promote understanding* (pp. 159-184). Mahwah, NJ: Lawrence Erlbaum Associates.

152. Shew, J. A. (1996). *Students' beliefs about mathematics and the way it should be learned: A story of struggle and change.* Unpublished doctoral dissertation, University of Wisconsin–Madison.

153. SIMMS Integrated Mathematics. (1993). *The SIMMS project, monograph 1: Philosophy statements.* Unpublished manuscript.

154. SIMMS Integrated Mathematics. (1997). *The SIMMS project: Final report.* Unpublished manuscript.

155. SIMMS Integrated Mathematics. (1997). *The SIMMS project, monograph 3: Final report.* Unpublished manuscript.

156. SIMMS Integrated Mathematics. (1998). *The SIMMS project, monograph 4: Assessment.* Unpublished manuscript.

157. SIMMS Integrated Mathematics. (1998). *The SIMMS project, monograph 5: The classroom.* Unpublished manuscript.

158. SIMMS Integrated Mathematics. (undated). *SIMMS integrated mathematics: A modeling approach using technology brochure.* Boston: Pearson Custom Publishing.

159. Simon, A. N. (1997). *Students' understanding of the comparison of the linear, quadratic and exponential functions.* Unpublished doctoral dissertation, University of Wisconsin–Madison.

160. Smith, J., and Berk, D. (2001, April 10-14). *The "Navigating Mathematical Transitions Project": Background, conceptual frame, and methodology.* Paper presented at the Annual Meeting of the American Educational Research Association, Seattle, WA.

161. Smith, J. P., Herbel-Eisenmann, B., Jansen, A., and Star, J. (2000, April 24-28). *Studying mathematical transitions: How do students navigate fundamental changes in curriculum and pedagogy?* Paper presented at the Annual Meeting of the American Educational Research Association, New Orleans, LA.

162. Smith, J. P., Herbel-Eisenmann, B., and Star, J. (1999). *Middle school students' algebra learning: Understanding linear relationships in context.* NCTM Research Pre-session of the Annual Meeting of the

National Council of Teachers of Mathematics. Reston, VA: NCTM. Unpublished document.

163. Smith, J. P., and Phillips, E. A. (1997). *Problem-centered algebra in middle school access via a broader set of skills.* Unpublished manuscript.

164. Smith, J. P., Phillips, E. A., and Herbel-Eisenmann, B. (1998). *Middle school students' algebraic reasoning: New skills and understanding from a reform curriculum. Proceedings from the 20th Annual Meeting of the Psychology of Mathematics Education, North American Chapter, Raleigh, NC, October 1998* (pp. 173-178).

165. Smith, M. E. (2000). *Classroom assessment and evaluation: A case study of practices in transition.* Unpublished doctoral dissertation, University of Wisconsin–Madison.

166. Smith, M. E. (in press). Practices in transition: A case study of classroom assessment. In T. A. Romberg (Ed.), *Insight stories: Assessing middle school mathematics.* New York: Teachers College Press.

167. Snipes, J., and Doolittle, F., and. H. C. (2002). *Foundations for success: Case studies of how urban school systems improve student achievement.* MDRC for the Council of the Great City Schools.

168. Spence, M. S. (1997). *Psychologizing algebra: Case studies of knowing in the moment.* Unpublished doctoral dissertation, University of Wisconsin–Madison.

169. Spencer, D. A. (2001). *Students' performance in mathematics.* Madison School District, Phoenix, AZ. Unpublished manuscript.

170. St. John, M., Heenan, B., Houghton, N., and Tambe, P. (2001). *The NSF implementation and dissemination centers: An analytic framework.* Inverness, CA: Inverness Research Associates.

171. Swafford, J. O., and Kepner, H. S. (1978). *A report of the evaluation of algebra through applications.* Unpublished manuscript.

172. Swafford, J. O. and Kepner, H. S. (1980). The evaluation of an application-oriented first-year algebra program. *Journal for Research in Mathematics Education, 11,* 190-201.

173. Swann, J. M. (1995). *Transition Math and PSAT scores.* Internal memorandum of February 15, 1995, to Michael Turner. Unpublished document.

174. Tetley, L. (1998). Implementing change: Rewards and challenges. *Mathematics Teaching in the Middle School, 4*(3), 160-165.

175. Thompson, D. R., and Senk, S. L. (1993). Assessing reasoning and proof in high school. In N. L. Webb and A. F. Coxford (Eds.), *Assessment in the mathematics classroom 1993 yearbook* (pp. 167-176). Reston, VA: National Council of Teachers of Mathematics.

176. True, G. N. (undated). *The effect of continuous distributed review on mathematics achievement.* Unpublished manuscript.

177. Truitt, B. A. (1998). *How teachers implement the instructional model in a reformed high school mathematics classroom*. Unpublished doctoral dissertation, University of Iowa.

178. University of Chicago School Mathematics Project. (1990). *UCSMP advanced algebra second edition: Summary of evaluation from teacher's edition*. Chicago: Author.

179. University of Chicago School Mathematics Project. (1990). *UCSMP algebra second edition: Summary of evaluation from teacher's edition*. Chicago: Author.

180. University of Chicago School Mathematics Project. (1990). *UCSMP geometry second edition: Summary of evaluation from teacher's edition*. Chicago: Author.

181. University of Chicago School Mathematics Project. (1990). *UCSMP spring newsletter, no. 7*. Chicago: Author.

182. University of Chicago School Mathematics Project. (1990). *UCSMP Transition Math: Summary of evaluation from teacher's edition*. Chicago: Author.

183. University of Chicago School Mathematics Project. (1992). *UCSMP functions, statistics and trigonometry: Summary of evaluation from teacher's edition*. Chicago: Author.

184. University of Chicago School Mathematics Project. (1992). *UCSMP precalculus and discrete mathematics: Summary of evaluation from teacher's edition*. Chicago: Author.

185. University of Chicago School Mathematics Project. (1996). *UCSMPerspectives spring newsletter, no. 13*. Chicago: Author.

186. University of Chicago School Mathematics Project. (1996-1997). *UCSMP winter newsletter, no. 20*. Chicago: Author.

187. University of Chicago School Mathematics Project. (1998). *UCSMPerspectives spring newsletter, no. 15*. Chicago: Author.

188. University of Chicago School Mathematics Project. (1999). 1999 August inservice evaluations. *In University of Chicago School Mathematics Project Conference Proceedings, Chicago, IL, August 2-3*.

189. University of Chicago School Mathematics Project. (1999). Program for 11th annual secondary inservice conference. *In University of Chicago School Mathematics Project Conference Proceedings, Chicago, IL, August 2-3*.

190. University of Chicago School Mathematics Project. (2001). Seventeenth annual secondary conference program booklet. *In University of Chicago School Mathematics Project Conference Proceedings, Chicago, IL, November 10-11*.

191. University of Chicago School Mathematics Project. (2001). *The University of Chicago School Mathematics Project 2000-01 descriptive brochure*. Chicago: Author.

192. University of Chicago School Mathematics Project. (2001). 2001 November secondary conference evaluations. *In University of Chicago School Mathematics Project, Chicago, IL, November 10-11.*

193. University of Chicago School Mathematics Project. (2002-2003). *UCSMP winter-spring newsletter, no. 30.* Chicago: Author.

194. University of Chicago School Mathematics Project. (undated). *Study skills handbook.* Chicago: Author.

195. University of Chicago School Mathematics Project. (undated). *UCSMP research and development.* Chicago: Author.

196. Usiskin, Z. (1969). *The effects of teaching Euclidean geometry via transformations on student attitudes and achievement in tenth-grade geometry.* Unpublished doctoral dissertation, University of Michigan.

197. Usiskin, Z. (1972). The effects of teaching Euclidean geometry via transformations on student attitudes and achievement in tenth-grade geometry. *Journal for Research in Mathematics Education, 3,* 249-259.

198. Usiskin, Z. (1982). *Van Hiele levels and achievement in secondary school geometry.* Paper presented at American Educational Research Association, New York, NY.

199. Usiskin, Z. (1997). *The evaluation of new curricula.* Unpublished manuscript.

200. Usiskin, Z. (in press). *A personal history of the UCSMP secondary school curriculum 1960-1999.* In G. M. A. Stanic and J. Kilpatrick (Eds.), *A history of mathematics education.* Reston, VA: National Council of Teachers of Mathematics.

201. Usiskin, Z., and Bernhold, J. (1973). *Three reports on a study of intermediate mathematics.* Unpublished manuscript.

202. Usiskin, Z., Senk, S., Hynes, C., and Siegel, C. (1990). *Report on the University of Chicago School Mathematics Project 1989 secondary summer institute.* Chicago: University of Chicago School Mathematics Project.

203. van Amerom, B. (2002). *Reinvention of early algebra.* Utrecht, Netherlands: CD–B Press, Center for Science and Mathematics Education.

204. van den Heuvel-Panhuizen, M. (1996). *Assessment and realistic mathematics education.* Unpublished doctoral dissertation, Universiteit Utrecht, Netherlands.

205. van den Heuvel-Panhuizen, M. (1996). *Developing assessment on problems on percentage: An example of developmental research on assessment problems conducted within the MiC project along the lines of Realistic Mathematics Education.* Unpublished doctoral dissertation, University of Wisconsin–Madison.

206. van den Heuvel-Panhuizen, M. (in press). Developing assessment on problems on percentage. In T. A. Romberg (Ed.), *Insight stories: Assessing middle school mathematics*. New York: Teachers College Press.

207. van Reeuwijk, M. (1993). *Assessment tasks designed to improve learning of mathematics*. Paper presented at the Annual Meeting of the American Educational Research Association, Atlanta, GA.

208. van Reeuwijk, M., and Wijers, M. (1997). Students' construction of formulas in context. *Mathematics Teaching in the Middle School, 2*(4), 230-236.

209. Van Zoest, L. R., and Ritsema, B. E. (1998). Fulfilling the call for mathematics education reform. *NCSM Journal of Mathematics Education Leadership, 1*(4), 5-15.

210. Verkaik, M. (undated). *CPMP student performance*. Holland Christian High School, Holland, MI. Unpublished manuscript.

211. Webb, D. C. (in press). Enriching assessment opportunities through classroom discourse. In T. A. Romberg (Ed.), *Insight stories: Assessing middle school mathematics*. New York: Teachers College Press.

212. Webb, D. C., Ford, M. J., Burrill, J., Romberg, T. A., and Kwako, J. (2001). *NCISLA middle school design collaborative third year student achievement data technical report*. Madison, WI: National Center for Improving Student Learning and Achievement in Mathematics and Science, Wisconsin Center for Education Research.

213. Webb, D. C., and Meyer, M. R. (2001). *Summary report of student achievement data for Mathematics in Context: A connected curriculum for grades 5-8*. Madison: University of Wisconsin–Madison, Wisconsin Center for Education Research.

214. Webb, D. C., Romberg, T. A., Ford, M. J., Kwako, J., and Reif, J. (2001). *NCISLA middle school design collaborative second year student achievement data technical report*. Madison, WI: National Center for Improving Student Learning and Achievement in Mathematics and Science, Wisconsin Center for Education Research.

215. Wijers, M. (in press). Analysis of an end-of-unit test. In T. A. Romberg (Ed.), *Insight stories: Assessing middle school mathematics*. New York: Teachers College Press.

216. Williams, D. (1986). *The incremental method of teaching algebra 1*. Unpublished research paper for University of Missouri, Kansas City course ED-621.

217. Wilson, M. R., and Lloyd, G. M. (1995). *High school teachers' experiences in a student-centered mathematics curriculum*. In D. T. Owens, M. K. Reed, and G. M. Millsaps (Eds.), *Proceedings of the Seventeenth Annual Meeting of the North American Chapter of the International Group for the Psychology of Mathematics Education*

(Vol. 2, pp. 162-167). Columbus, OH: The ERIC Clearinghouse for Science, Mathematics, and Environmental Education.

218. Winking, D. (1997). *The Connected Mathematics Project: Helping Minneapolis middle school students "Beat the Odds," year one evaluation report.* Unpublished manuscript.

219. Winking, D. (1998). *The Minneapolis Connected Mathematics Project: Year two evaluation report.* Unpublished manuscript.

220. Wolff, E. (1997). *Summary of matched-sample Stanford 9 analysis comparing IMP and traditional students at Central High School, Philadelphia, PA.* Unpublished manuscript.

221. Wolff, E., and Decktor, P. (1997). *Summary of matched-sample analysis comparing IMP and traditional students at the Philadelphia High School for Girls on mathematics portion of Stanford 9 test.* Unpublished manuscript.

222. Zawojewski, J., Robinson, M., and Hoover, M. (1999). Reflections on developing formal mathematics and the Connected Mathematics Project. *Mathematics Teaching in the Middle School, 4(5),* 324-330.

223. Ziebarth, S., Slezak, J., Lagrange, D., and Kleinfelter, N. (1997). *Teaching a reformed high school mathematics curriculum: Inservice and preservice perspectives.* Paper presented at the Annual Meeting of the Association of Mathematics Teacher Educators, Washington, DC.

224. Ziebarth, S. W. (1998). Iowa Core-Plus Mathematics Project is evaluated as a success. *ICTM Journal, 26,* 12-21.

225. Zucker, A. A., and Shields, P. M. (1995). *Evaluations of the National Science Foundation's statewide systemic initiatives (SSI) program second-year case studies: Connecticut, Delaware, and Montana.* Arlington, VA: National Science Foundation.

TABLE B-1 Distribution of Background Information and Informative Studies by Curricula

	Number of Studies
NSF-Supported Curriculum Name	202
Everyday Mathematics	16
Investigations in Number, Data and Space	9
Math Trailblazers	6
Connected Mathematics Project (CMP)	42
Mathematics in Context (MiC)	52
Math Thematics (STEM)	13
MathScape	5
MS Mathematics Through Applications Project (MMAP)	7
Interactive Mathematics Project (IMP)	12
Mathematics: Modeling Our World (MMOW/ARISE)	5
Contemporary Mathematics in Context (Core-Plus)	19
Math Connections	6
SIMMS	10
Commercially Generated Curriculum Name	73
Addison Wesley/Scott Foresman	1
Harcourt Brace	0
Glencoe/McGraw/Hill	4
Saxon	21
Houghton Mifflin - McDougal Littell	1
Prentice Hall/UCSMP	46
Number of evaluation studies	225
Number of times each curricula are in each type	275

Appendix
C

Outcome Measures

American College Test (ACT)
California Achievement Test (CAT)
Comprehensive Test of Basic Skills (CTBS)
Fennema-Sherman Mathematics Attitudes Scales
High School Subject Test (HSST)
Horizon Research, Inc. Teacher Survey
Illinois Goal Assessment Program (IGAP)
Illinois State Achievement Test (ISAT)
Iowa Algebra Aptitude Test (IAAT)
Iowa Test of Basic Skills (ITBS)
Iowa Tests of Educational Development - Quantitative (ITED-Q)
Massachusetts Comprehensive Assessment System (MCAS)
Massachusetts Educational Assessment Program (MEAP)
Mathematical Aptitude Test version 6 (MAT-6)
Mathematical Aptitude Test version 7 (MAT-7)
Mathematical Problem Solving Test (MPST)
Missouri Assessment Program (MAP)
Missouri Mastery and Achievement Test (MMAT)
National Assessment of Educational Progress (NAEP)
New Standards Mathematics Reference Examination (NSMRE)
Northwest Achievement Levels Test (NALT)
Orleans-Hanna Algebra Prognosis Test
Otis Lennon School Ability Test (OLSAT)
Preliminary Scholastic Achievement Test (PSAT)

Quantitative Reasoning Test (QRT)
Scholastic Aptitude Test (SAT)
Science Research Associates Test (SRA)
Second International Mathematics Study (SIMS)
Stanford Achievement Test 7th edition (SAT-7)
Stanford Achievement Test 8th edition (SAT-8)
Stanford Achievement Test 9th edition (SAT-9)
Texas Assessment of Academic Skills (TAAS)
Third International Mathematics and Science Study (TIMSS)
Washington Assessment of Student Learning (WASL)

Index